Vending Machines

A Simple Guide to Start a Vending Machine

(The Ultimate Complete Guide to Automate Your Income and Achieve Financial Transformation)

Thomas Shouse

Published By **Phil Dawson**

Thomas Shouse

Vending Machines: A Simple Guide to Start a Vending Machine (The Ultimate Complete Guide to Automate Your Income and Achieve Financial Transformation)

ISBN 978-0-9959231-7-1

Legal & Disclaimer

Table Of Contents

Chapter 1: Where To Put Your Vending Machine

Starting this business agency is just like asking which came first, the hen or the egg. The reason why is due to the truth you may ask a comparable query right proper right here, as in must you purchase a device or find out a vicinity first?

There are benefits and drawbacks to doing each one in advance than the alternative. So preserve in thoughts that you can do matters in each order, however I want to cowl the basis of this so you could make a knowledgeable choice.

Why You Should Buy a Machine First

One of the benefits to getting your device first is that you'll make certain your tool is ready to move at the same time as you do have an area. You don't need to keep a place prepared as a manner to get your merchandising tool and if you're expecting your vending device to be shipped to you, there is the far flung possibility that a few factor may also moreover need to transport wrong alongside transport delays, damage in a few unspecified time within the future of transit, or it not functioning well.

You can also want to ensure plenty of time to repair capability troubles like this if you wait to discover a area. You can get your insurance facts in advance as well, and this could assist to make the system as seamless as feasible in phrases of securing a place to your system. This isn't to say that obtaining your tool first is without its flaws.

Finding vicinity can from time to time be greater of a challenge counting on in which you stay. You may also moreover want to offer yourself as lots time earlier as you can to make certain you're going with a purpose to regular an awesome area within the first place. If you've got a system and no location to place it, it will in all likelihood live stuck on your garage or a garage unit till you discover an area.

Then you moreover may additionally want to transport it from your own home or storage unit to the area wherein it's going to are dwelling and a good way to value you a exceptional little bit of money until you have got got the talents to deal with that for your very very personal.

Why You Should Find a Location First

By finding an area first, you'll ensure that your system has a domestic earlier than you even buy it. You also can bypass the device without delay to in which it'll stay so there's no double transferring.

The biggest drawback although is the truth that the area ought to get purchasers remorse so to speak, or they located a totally precise machine in area of yours because it's prepared to go. If this takes area to you, it can be quite devastating, so which component need to you deal with first?

Personally, I but take delivery of as authentic with that it makes the maximum experience to try and discover a place first in advance than you purchase a tool. There are of course elements that would alternate this. For example, you might be shopping for your machine from foreign places, and you could be waiting up to sixeight weeks for your device to obtain.

In this situation with the timeframe being goodbye earlier than you'd have a tool to offer, you will be more likely to get burned, so that you'd be better off prepared until it's approximately to reach. So with that during mind, allow's flow earlier and speak approximately securing your first region.

What's Convenient for You

We don't need to rush out and start contacting agencies right out of the gate. Not all locations are created similarly, and the number one component you need to consider is yourself. I accept as true with that this is greater important than some component else due to the fact deciding on the wrong location can purpose you to resent your business enterprise.

For instance, you need to reflect onconsideration on your contemporary time table and give you what the precise scenario might be for a location to mesh together with your schedule. Let's say as an example you determine entiretime from 8 a.M to 5 p.M. If this is the case, then it's no longer going to make feel to area your system in a corporation whose jogging hours are similar to your work hours.

You'll by no means have the time to restock your gadget. This area may be highquality profitable, however seeking to make it art work is actually going to strain you out if you're now not going a terrific manner to constantly and reliably advantage get right of entry to on your machine. If your paintings schedule became from noon to 8, 2 to ten, or in a unmarried day, then you have a big window of time as a way to get entry to your tool.

You moreover need to reflect onconsideration on how a ways away your gadget is located far from you as you may have to component in that element. These issues can of course be solved via locating something that's open longer, which include a mall or dorm corridor. So first hassle is to take into account on the equal time as you'll be to be had to restock your system, and then use that to higher help you decide what varieties of businesses you want to target.

What Are Some Spots that You Should Consider?

Any workplace might be an awesome one so hone in on genuinely those thoughts and forget about the whole thing else. Any workplace whose number one target audience is ladies may be a exquisite touchdown spot to your tool, so don't robotically rule a few element out. With that out of the way, right here are some suitable locations to reflect onconsideration on:

Rest Stops and Travel Stops

The one number one drawback of rest stops is they may be out of the manner probable. If the closest rest prevent to you is simply too a long manner away, then it's satisfactory to keep off until you have got different locations set up. This way you'll be capable of construct a direction that's closer to that location.

Aside from that, relaxation stops can be a extremely good place which will strive out. People who prevent at rest stops have normally been travelling for a while and that they want a damage. A regular vending machine is an apparent choice, however so is a splendor vending machine due to the reality humans run out of their resources after they adventure.

They underestimate how a splendid deal they need to p.C. Or they forget about to p.C. Something, and that they'll be pleasantly surprised when they encounter your device. You additionally have boredom running for your select right here too. People will stay at a rest prevent for a chunk of time earlier than they hit the street another time and that they'll need some component to do, so why now not test out what this fancy merchandising tool has to provide?

College Dorms

College dorms are incredible for a plethora of motives. For one trouble, they're open 24 hours an afternoon due to the fact college students want access to wherein they stay glaringly. Dorm rooms furthermore have more youthful women, that is a part of your

8

aim market, in order that's a plus right there.

You recognize that your tool may be in the front of folks that may be interested in what you're supplying. Lastly, there's now not as lots opposition for beauty vending machines due to the reality most usually have a tendency to intention places like department shops, so you ought to have an less difficult time getting your foot inside the door. The one downside to college dorms is that some college kids will pass domestic in the course of the weekend so foot visitors may be lower at the weekends.

Airports

It doesn't take a genius to recognize why an airport is a golden possibility to your vending device. They're open 24/7 and function splendid foot internet web page site visitors all of the time.

The handiest real drawback is that there's realistically only going to be one or perhaps airports to your area that you could aim. So you obtained't be capable of churn out an entire lot of machines from this kind of place, but even securing definitely one location is big.

Hotels

Open 24/7 with foot site visitors continuously, in reality not as tons as an airport. Hotels are terrific due to the truth they're going to be an entire lot of them in a unmarried region. This offers you the chance to hit up pretty some distinctive hotels and when you land with one, you can extra without trouble increase within the course of others if you're handling a series.

Work Offices

Work offices can be another underrated area that allows you to area your tool at because of the truth there isn't going to be a whole lot of opposition. The key's to select out out a commercial enterprise business enterprise that employs hundreds of people. Smaller organizations in fact received't generate heaps foot website site visitors surely in order that they received't be actually well really worth a while.

You additionally want to consider the fact that maximum company places of work are closed at the weekends because of this that that your device gained't be making any money at the weekends. Due to those drawbacks, I might appearance elsewhere first after which begin searching at organisation offices in case you're now not able to discover some thing else.

Malls

Putting your gadget in a mall is a famous desire for a purpose. Lots of foot site visitors, plenty of your target marketplace in a unmarried vicinity, and people are within the mindset to buy. You'll certainly want to hit up each mall to your place and see if any availabilities are open.

Even within the event that they aren't, you'll want to have a observe up each 6 months to look if something has unfold out. You want to be as persistent as need be to make sure you're placing your machines inside the extraordinary places feasible. And following up is some thing you want to do with any place you want get right of entry to to.

You won't get in in your first attempt to that's right enough. Start planning for putting your 2nd tool there or probable transferring your first tool in case you're capable of due to the reality the brand new area may be greater worthwhile.

Something you'll take a look at about plenty of these suggestions is that they have got lengthy open hours and constant foot site visitors. This is the proper combination you need to search for, however all once more it doesn't suggest you need to rule out different places.

Now that you have some mind for locations you may aim, you don't want to run in blindly. You need to do some research to get a few statistics to your aspect. This will assist with figuring out a region's profitability and help you evaluate placement costs.

Conducting Proper Location Research

What I'm approximately to indicate to you isn't precisely going to be fun. But what I can inform you is that it is going to be well worth the strive due to the fact no character else is probably going to be doing this. Essentially what you're going to do is flow to three of the only of a kind places and measure the foot site visitors.

This will will will assist you to apprehend when you have a vicinity that's surely well well worth it or now not. You'll want to shop for a clicker for this or download an app so one can assist you to depend the range of human beings that stroll thru. Once

13

you have that, all you want to do is sit down on the capability landing spot in your gadget and depend the amount of humans that walk with the aid of for one hour.

This will provide you with a first rate idea of methods busy an area is. To maintain this constant, you'll need to try to skip at pinnacle instances for each administrative center. For instance, your results will range considerably if you go to the mall on a Monday morning in preference to a Friday night time time.

The equal thing would possibly follow if you desired to diploma the foot web web page visitors in an airport. If you degree foot web site visitors inside the airport on a Wednesday afternoon, your results is probably manner unique as compared to site traffic on a Friday afternoon. But keep in mind comparing website visitors from the mall on a Monday morning in comparison to Friday afternoon site traffic.

Chapter 2: Buying Your First Vending Machine

Since I've already covered finding a location, now it's time to speak approximately shopping for your first vending device. There are quite some unique options that you have to be had to pick out out from, so you'll need to ensure you gradual down and suppose this purchase via due to the fact vending machines aren't reasonablypriced.

What Type of Vending Machine Do You Want?

Back in the day, merchandising machines surely had the identical antique buttons which you can press to get your item and that they best normal coins. Nowadays topics are a piece specific as technology has superior. A lot of merchandising machines will take delivery of credit score score playing cards and some fancier fashions absolutely have contact monitors to allow clients to look what's to be had and make their selection from there.

You can also purchase any sort of vending device which you could be interested by buying as a present day or used circumstance. Therefore, the number one problem you want to decide on is the sort of gadget that you need. Do you need a device which can gather credit rating gambling gambling cards?

Do you need a present day device with a touch show show display? You also should keep in mind the price proper right here. You need to get a used system for under $1,000, but some topoftheline machines can rate you as tons as $five,000.

There's virtually a large range from what you could purchase. So fee is some problem you need to think about because it will determine how long it takes you to make a income. If you purchase a $5,000 system and also you need to pay $500 regular with month in hire, it's going to take some time before you are making your first greenback.

But you need to recollect matters with an extendedterm attitude. This device might be extra appealing for humans to need to shop for from and it will be a better look for your brand. You also can fee a bit more because of the advanced revel in and better perceived price.

In the long run, it will pay off. However, you may now not have the capital to make investments in the wonderful vending device and that's good enough. It is probably higher to buy what you could offer you with the cash for and get began in area of to wait to try to store sufficient coins to shop for a fancier device.

As time is going on, you could constantly sell off your initial device and update it with a fancier one when you start to make extra cash. You additionally want to consider consistency if you need to very private more than one system.

You want to create a similar brand image with every device in order that your machines turns into extra recognizable as time is going on. That's a few factor you may awareness on in some time at the same time as you're prepared for your 2d device, however for now, permit's communicate greater inintensity about your first device.

Should You Buy a New Machine or a Used Machine?

The obvious pro to looking for a used gadget is that it is going to be much less highlypriced and if that is what you can manage to pay for, then no issues. You should purchase a used device and do extraordinary with it. A new device will include a warranty in maximum times relying on what supplier you purchase it from.

A new device can also be in incredible condition, at the equal time as a used device can also need to have dents, scratches, or different beauty issues that could want to be repaired earlier than you have humans buy from it. A used machine could not be in strolling order, but that's a entice you acquired't ought to worry approximately after analyzing this chapter.

Either way, the choice is as lots as you. If you have the extra money, trying to find new is in fact much much less hard and may provide you with higher peace of thoughts because you don't must fear about overlooking a few component. So I'll first cowl shopping for a brand new tool and then used.

Where to Buy a New Machine?

You should purchase a current device domestically or distant places. The benefit to purchasing for distant places is that you could usually get a higher deal because of the reality that is in which most merchandising machines are being synthetic within the first area. The disadvantage to buying foreign places is that it's going to take lots longer so that it will gather your tool.

Additionally, because of the tool being shipped from farther away and it changing fingers at multiple points, there's extra of a chance for the machine to arrive damaged. This usually obtained't take area, but it is more likely for some factor to be broken the farther it travels. Also, if your machine does arrive broken, it is going to be a chunk bit greater tough to get the trouble regular and it's going to take longer clearly because of logistics.

When you purchase domestically, you can count on to pay extra, but the technique can be smoother common. What I advise which you do is buy community if viable. You can do this with the useful resource of looking "vending tool dealer near me."

This will give you some right alternatives that you may begin to examine. If you're no longer capable of discover whatever inside an cheap distance, you could continuously do a stylish are searching for and discover a dealer inside the US an top notch way to deliver your system to in which you want to. The advantage of getting a issuer near you is that you can check out the tremendous machines they have to provide in character.

Whatever device you emerge as shopping for gained't must journey a totally a long manner distance both, that could be a big plus. When searching at new machines, one large issue that you want to ensure you get is a guarantee. Some carriers will encompass this, and different places will provide this as a separate buy.

If your machine doesn't consist of a guarantee, I trust it's in fact nicely well worth the charge. Imagine some element going incorrect together with your tool rapid once you purchase it and you haven't any idea the manner to move approximately fixing it. If you've got sufficient coins saved as an awful lot as pay a person to restore it, then that is no huge deal. Having a assurance can come up with peace of mind simply in case a few trouble does cross incorrect.

Where and How to Buy a Used Machine

Buying a used gadget will variety significantly in evaluation to purchasing for a modernday device. For starters, you do have a chunk greater range for wherein you could keep. You can search on line marketplaces collectively with Facebook Marketplace and look for people promoting their vending machines.

The cool aspect approximately attempting to find a device on Facebook marketplace or some thing comparable is that you may discover a good buy in case you're affected person enough. This is typically going to be the most inexpensive way to shop for a merchandising gadget. There are functionality downsides that you need to look out for even though.

Buying used from an established organisation is specific due to the fact their recognition is on the street and they don't want to get keep of a horrible assessment. This will help to guard you from shopping for a few thing that doesn't art work. With a web marketplace, this doesn't exist due to the fact the vendor might not care.

They have a large object that they need to put off. They understand they'll have lots of hobby if the rate is aggressive sufficient. Once the sale is made, the seller can wipe their palms smooth.

Then even as you attempt to use the tool, you word that a few detail of it's far damaged and also you experience ripped off. You can not anticipate that the vendor indexed each sickness with the device. Instead, you want to expect the opportunity.

Assume that there is probably some thing incorrect with the device until you're capable of verify that it completely works. This is mainly the case if you see a price that looks too properly to be genuine. In the case of merchandising machines, they're cumbersome devices which can be tough to move.

Someone may be in a scenario in which they need to make a fast sale, so don't mechanically expect the device doesn't artwork. It's actually honestly well worth going and locating out because of the reality you can end up getting a bargain. One drawback to searching out used from an man or woman dealer is that you obtained't have a guarantee, but, the rate difference can very well make up for this.

So what need to you look for to assist shield you from getting ripped off? The first thing is the most apparent and that's to plug the device in. Like I said earlier, merchandising machines are bulky and if the machine is in a vicinity a long way a ways from an electrical outlet, it can now not be sensible to move the gadget inside obtain of the hollow.

This is why you need to deliver an extension twine with you to be organized. Don't expect it will probable be close to an outlet or that they'll have an extension wire for you. When you plug it in, you want to ensure it lighting up.

Then check out the coins slot and ensure that it accepts bills and alternate. After that, you'll need to test out every button and make certain that each one can be pressed and is functioning. Also, test everybody slot wherein a product may be placed and make certain that it could flow into well on the identical time as that slot is selected.

If you're purchasing for a machine with a touch show show screen, make certain there aren't any lifeless pixels at the display display. So for example while the show display lighting fixtures up, a vain pixel will usually be black no matter what colour is meant to be showing. Additionally, check the show to make sure that it's responsive and that it runs with out troubles.

You don't need to have the device lagging within the returned of from people's inputs. If you do all of these items, you may enjoy assured that the gadget works, however you moreover may have to bear in mind aesthetics. Is there any beauty damage that's large at the same time as the device is being used?

Damage on the decrease once more of the system isn't a huge deal because of the truth that obtained't be seen. Cosmetic damage to the element or the the front of the tool can be located. Some sellers will thing out the ones types of damages to you, and others received't.

It is your obligation to research the device for scratches and dents earlier than you buy it. If you word some splendor harm that wasn't referred to inside the description, then you can component this out and use it as leverage to help negotiate a lower fee. If the cultured harm is simply too fantastic, then it's probable higher to skip on it, than to pay to get it consistent.

Once yet again, the vital thing even though is to in no way anticipate that every one of the damage is probably indexed within the description. If you're able to do those objects, you then clearly'll set your self up for fulfillment. The exclusive way to shop for used is just like purchasing for new, and that's to buy from a supplier.

I although advise testing out the tool if feasible earlier than you buy. If you're unable to check out the tool, you then definitely sincerely'll want to make certain the company gives a assure that the device will art work, in any other case, you're better off going to a tremendous seller.

One Other Point to Consider

Lastly, while it involves searching for used, there's a way that you can kill birds with one stone doubtlessly. The way you may do this is with the resource of buying a device this is already in a specific location. This will will let you no longer should cope with the stress of locating and securing an area, however you continue to need to do your research. You don't want to take over a tool that isn't doing properly.

You also have to take into account the truth that you'll likely be taking on a snack tool and converting it right right into a splendor tool. So genuinely due to the fact a snack gadget did correct or awful in a highquality vicinity does no longer advocate that the same trouble may be the case with a splendor merchandising device. If you come across a exceptional area from a web supplier, although, this may be well worth exploring to help get topics off to a quicker begin.

Getting Your Machine Wrapped

You might be questioning, "Wait don't I want to shop for a system that's already constructed or used as a beauty vending machine?" No, you don't, as an opportunity you need to interest at the tool itself and ensure the system is in operating scenario and that it's the fashion you want. If all of that tests out, you can rent a business printing commercial enterprise company to wrap your device with something layout you want.

You can use a format that great represents your logo. Some printing corporations will help to create a format for you, whilst others will excellent do the printing and wrapping. Therefore, you can want to rent extra assist for the introduction of the design in case you're not able to do this yourself.

Chapter 3: Negotiating And Getting The Deal Done For A Location

Finding a capacity touchdown spot doesn't propose hundreds, if something, if you're now not capable of seal the deal. Another place also can seem top, but then the rent ultimately finally ends up being plenty extra than what you bargained for. In this economic catastrophe, I need to assist prepare you simply so you will be prepared to barter and get the exceptional deal possible for your machine.

Do Your Research First

Before you pass spherical contacting all and sundry to doubtlessly region your vending gadget, you want to do your studies first. This method finding functionality landing spots and actually going to the ones places and measuring foot internet page visitors. This is prime as it will come up with the facts you want for while you do begin speaking to the character in rate of the building.

The factor is that it could be clean to bypass this step and go right away into getting costs. This is a mistake, so begin by way of getting a list of potential places and then skip from there.

Be Patient

One way to get the short forestall of the stay with a deal is to be keen. If you found that this is your one chance and it'll be lengthy past day after today, you'll be much more likely to make an impulse choice, that's what can cause you to pay more regular with month than you in reality must be. Instead, I need you to recognize that gives will come and bypass.

If you pass over out on a deal that regarded exquisite, don't lose sleep over it because of the truth a new area will are available your radar at some point if you preserve looking. The trouble for max humans is they're now not affected person, simply so they're going to take the number one deal that they can. This gives you little leverage for negotiation.

Instead, endurance can prevent a ton of cash due to the truth do not forget in case you're able to wait a chunk and you emerge as with an area that's open more hours, has more foot web page visitors, and is in a similar charge variety as a competitor you're looking at. Wouldn't you be satisfied which you checked out more than one options as they got here up in vicinity of leaping at the first possibility? Who is privy to, you may not additionally be up towards any opposition, so you'll be bidding in competition to yourself!

Don't Act With Desperation

Similar to the issue above, on the equal time as you're speaking with various agencies, you don't need to come across with a splendid sense of urgency or desperation. This will tip off the alternative facet which you need this deal to undergo extra than they do. You'll make it heaps harder on yourself to get everywhere whilst the fact is that there's no need to act determined due to the reality there are typically other options available.

A few ways that you may encounter as unneedy is to do your studies. If you recognize the amount of website online visitors that comes thru the place, then you definately'll be able to speak with self assurance as to why the numbers don't make enjoy.

Price is Negotiable

Another key trouble to maintain in mind is that you can negotiate price. Some places can be greater hard to budge than others. Typically, bettername for locations is probably extra difficult to sway than locations that aren't going to have as masses name for.

Either manner even though, you'll regardless of the fact that want to put your remarkable foot ahead due to the fact there's no damage in attempting. Often instances we're knowledgeable the price of a few issue and we don't count on instances about it and we pay it. But whenever a commercial business enterprise gives you a price, you don't want to move away it at that. You can try to negotiate some element.

You is probably capable of lower the charge paid on profits or remove it truely. Or you might be capable of advantage some super incredible perk from the organisation. For instance, possibly your device is going in a nail salon, and despite the fact that they're now not willing to adjust their price, they nonetheless decide to provide you one loose pedicure regular with month or offer you a reduction on distinct merchandise that they promote.

Go Under the Actual Amount You Want to Pay

With organizations which can be more established or which have had vending machines in their location in advance than, you'll likely be knowledgeable a price and subjects will development from there. Even in case you don't expect that a administrative center will budge on its fee, it's constantly definitely really really worth negotiating to look if you may get a better deal. The ultimate problem you need to do is take the number one deal you encounter, so that you need to try to negotiate at multiple places to make sure you're getting the superb deal possible.

One tip with reference to giving out your very very own price is to transport under the quantity you honestly want to pay. For instance, allow's say a administrative center tells you it will price $750 regular with month to save your merchandising device on their belongings. If you determined $six hundred is a sincere charge and that's what you need to pay or get as close to that range as feasible, you then definately don't want to counter with $600.

Instead, you'd be higher off countering with something like $500. The cause why you want to move decrease is due to the reality the proprietor will counteroffer and their new significant range might be much more likely to be spherical what you definitely desired. Typically human beings will offer the median when they counteroffer.

So within the case of an preliminary offer of $750 and also you countering with $500, the median might be about $625, this is near the range you virtually desired of $600. You should take a counteroffer into consideration because of the reality human beings want to revel in like they're those getting a bargain. That can't display up if they sincerely go along with the first amount that you throw accessible.

You might not enjoy comfortable countering with a number of that a bargain decrease than what the opposite aspect initially supplied, but if you're willing to push past that preliminary little little little bit of uncomfortableness, then you clearly offer your self the functionality to get a higher deal. The detail you need to take into account is that the worst someone can say isn't always any. So you exceptional have blessings to benefit without any ability loss.

Someone isn't going to raise their initial asking fee because of the truth you gave them a suggestion they didn't like. You moreover anchor yourself at the identical time as you deliver a lower range. So as an instance, allow's say you visit a automobile dealership and you see a $sixty five,000 vehicle next to a $30,000 car.

The $65k car makes the $30k automobile appear a whole lot extra inexpensive in spite of the fact that $30k is still an entire lot of money. The same detail happens in opposite basically on the same time as you call a low however lowcost fee. That will set the bar for the thing in which the opposite birthday celebration will counter from.

Don't Give Out the First Number if Possible

Another element to be privy to with reference to negotiating is doing all of your wonderful to avoid giving out the primary amount. Going 2nd gives you a large benefit due to the fact you haven't any idea what the other problem is thinking. Their initial variety might be way decrease than what you had been expecting to pay.

This is much more likely to take region in a enterprise agency that isn't as installed or hasn't had vending machines in advance than. For a place much like the mall or an airport, they'll be much more likely to recognize their numbers, so don't count on a few excellent deal. But for a place like a nail salon, they'll be much more likely to offer a good buy due to the fact they've likely never performed this before.

Going second for giving out various essentially lets in you to appearance the playing playing cards in your opponent's hand earlier than you have got to expose any of your very very own. You is probably on the same net web page and matters will improvement smoothly if that's the case. What you don't need to show up is you supply out some of and they were wondering a few thing an lousy lot higher from what you first of all provided.

So how do you move about looking to get the alternative side to provide a whole lot of first? Well, one of the fantastic matters you could do is ask. You can say, "So what did you've got in thoughts for the monthtomonth rent?" You'd be amazed at how powerful this may be.

If you're requested to offer out more than a few first, don't sense the pressure to provide out a range of. Instead, you could placed it yet again on the other person by means of way of saying a few issue like, "I'm not exquisite, what did you have got in thoughts?" Sometimes although you may not be capable of avoid tossing out the primary range.

So what do you do at the same time as you encounter someone who's stubborn and gained't budge? Well, this is in which it's essential to do your research so that you comprehend how wellknown of an area that is. Based on that knowhow, you can provide you with numerous that you're aiming to pay and then make certain to to start with provide much less than that.

From there the alternative facet can counteroffer, however the key to giving out the primary range is to purpose low. Once you call a fee, that's the component of no go back. For example, in case you begin via announcing you think $650 in keeping with month is straightforward, then you're now not going to end up with a few factor that's an entire lot a lot much less than $650 constant with month.

So don't supply out quite a few that you're going to be disappointed with. Instead, cross under what you clearly want, and opportunities are pinnacle that you'll arrive across the huge range you have got been hoping for first of all.

Try to Prioritize Negotiating Percentage of Sales Over Price Per Month

As I've stated in advance than, a few locations will try to price monthtomonth lease and further to that, they'll want a small percentage of the profits that your machines make every and every month. Yes, you may negotiate every the monthtomonth hire and the share of income, however one has to take precedence over the opportunity. If you're managing a place that desires a piece of the pie, I endorse through the usage of starting to negotiate down the proportion of sales first.

This may have a larger functionality on your not unusual earnings than the monthtomonth hire. Your hire will stay the same irrespective of what. However, in case you hit on a in fact particular place and sales are booming, then decreasing the percentage of income could have a bigger impact on the earnings which you preserve.

Of route, the place is probably useless and you might be better off searching for to lower the monthtomonth hire. You'll be capable of make the right selection on what to attention your negotiation on by way of way of referring lower back in your studies. If you recognize the area is a in reality accurate one with a number of visitors, then attempt to decrease the percentage of income that you supply them.

If foot web page site visitors isn't the amazing based totally on what you can acquire, then a better play possibly can be to try to lower the monthly rent first. If you're doubtful, then go along with the proportion of income.

Contact the Person in Charge

The first step to getting your merchandising device in a workplace is to touch the right character. This is simply much less tough than you would in all likelihood assume in maximum instances. So for example, in case you're searching for to located your merchandising tool in a mall, then do a web look for the mall's call, followed thru manner of "stylish supervisor."

This ought to give you the cellphone quantity of the individual that you need to touch. If you come upon a person who isn't the individual you need to speak to, it's no longer a massive deal due to the reality they'll be capable of direct you to the individual that you do need to get in touch with. Doing a brand new search for the place of business plus the general supervisor or belongings manager could be enough to influence you inside the proper route most of the time.

Once you do get ahold of the right man or woman, you'll sincerely united states of america who you're and what you're looking for to do. Then ask the man or woman what their fees are for the hire and you could improvement from there. If that is the first region you're contacting even though, thank them for the records and inform them you'll suppose on it and get again to them as quick as feasible.

Then you're going to begin repeating the manner for special places of business business enterprise. The difference is now you can use the first place's fees in opposition to different ability touchdown spots. So as an example, if you get a quote for $six hundred in line with month and no percentage of sales, then you can factor out that in case you get a quote from another region offering $sevenhundred a month.

You can tell them which you wouldn't be able to acquire $700 due to the fact you have were given already were given a wonderful spot that have become involved for $600. If they'd be capable of in shape the price, then you definitely'd be willing to seriously keep in mind it. This works properly for corporations that don't have a wonderful deal opposition for someone to location their vending machine of their space.

For extra installed locations of corporation, they may say the fee is what it's miles, which in that case no worries. Regardless of the response, that is why it's critical to do your research so you can create the capacity of the use of businesses closer to every different to help decrease the fee a few. One different aspect well well well worth noting is which you're going to ought to touch quite some businesses in advance than you encounter one which's involved or has the supply.

Chapter 4: Finding A Supplier For Your Beauty Products

To have a a achievement merchandising tool, you need to have items to promote. In a outstanding worldwide, you'd get the fantastic highquality products for the most inexpensive possible rate and that's how topics would possibly skip. Typically although, you need to find out the proper balance among rate and awesome.

You don't need topics to be too luxurious to in which humans can't discover the cash to your products, but on the equal time, you don't need lowawesome products that permits you to leave you with dissatisfied customers. Therefore, locating the proper company which can deliver you with satisfactory merchandise at a superb price goes to be key to being a achievement along with your merchandising gadget.

What Should You Stock Your Vending Machine With?

Before you go out seeking out the right dealer, you want to recognize what you're going to promote. The component approximately the beauty enterprise is that it is able to be as huge or as narrow as you want it to be. You ought to sell one unique form of product, collectively with lashes, lip gloss, nail polish, eyeliner, braids, mascara, or distinctive makeup resources.

The oneofakind option you have got were given is to promote a aggregate of severa items to help supply your customers extra of a selection to pick out out from. You also can bundle multiple gadgets collectively and promote that as a package deal. So for instance, you may package deal deal a lip gloss, more than one eyelashes, and nail polish multi function package that a person can purchase.

This can be an remarkable way to shop a few more region in your machine if you want to. You may additionally have a slot committed to thriller baggage. The patron gained't recognise what's contained of their bag until they buy it and this can be fun for the client and assist to pressure earnings because of the fact people like surprises.

There's no right or wrong manner to go approximately what you make a decision to put in your device. The cause for that is due to the truth you satisfactory have a lot area to your gadget for property. Just approximately any product you can take into account goes to vary in some manner or every different.

For example, there are unique patterns and lengths of eyelashes. There are firstrate sunglasses of nail polish. Offering much less products lets in you to move deeper with the precise products that you bring, so you need to offer greater colours, patterns, lengths, and lots of others. It's as tons as you to decide what form of a stability you want to strike.

For me in my opinion, I need to provide an superb kind of merchandise and stick with the most wellknown styles of that product. For instance, if I'm promoting nail polish, I'm pleasant going to offer the maximum not unusual solar shades and live some distance from area of interest colours. If you had countless location, this wouldn't be an hassle, but you need to be a chunk ruthless with what you hold and what you don't.

The accurate records is that you're now not stuck with what you make a decision to install there. You can usually change subjects up the manner you want as it's your machine. You can take out merchandise that aren't selling properly and try a few aspect specific in there.

What Price Should You Sell Your Products?

Pricing is a few other essential piece of the puzzle that you want to consider. By the man or woman of the industrial agency, you're going to have monthtomonth prices, so that you need to mark your merchandise up sufficient to make sure they will be capable of cover your running fees. You also want to observe some issue smooth.

You don't need to have a few complex set of pointers you need to study for each product which you need to sell. Instead, you need some thing easy with a purpose to artwork for any product which you intend to install your device. This is why I advocate that you carry out at a 50% earnings margin.

So if it expenses you $2.50 to buy a product, you may sell that product for $5. This permits to maintain matters smooth, that's what you need so you can interest your hobby on different topics. That could likely appear a bit steep in the beginning look, but don't forget your rent isn't reasonablypriced and also you're shopping for in bulk.

The client is best looking for what they want or they're seeking out for the experience. Much like the goods you promote, your pricing is fluid. If income are booming, you can boom your margin to fifty five or 60%.

If you're no longer transferring your splendor products the way you'd want to, then you could decrease that margin proper right down to forty% and spot if that entices greater human beings to shop for.

Where to Find Your Beauty Supply Vendor

Finding a supplier on your merchandise can be traumatic, but it actually boils right all the way down to trial and mistakes till you find out a organization that is a top match to your agency. To begin your are seeking, you can search for carriers on social media, together with Instagram, or through using the usage of the usage of an online are attempting to find engine. Either manner, the device may be very similar, so I suggest trying out both.

You'll need initially the aid of typing in a phrase together with beauty product issuer, beauty wholesaler, beauty product wholesaler, beauty product supplier, and so forth. These styles of phrases and their variations will offer you with a tremendous style of agencies that you could begin contacting. If you're selling handiest one product which incorporates lashes, you may search for some thing together with eyelash seller or eyelash wholesaler.

It's additionally crucial to recollect that in case you need to sell more than one diverse products, you don't need to get all of your products from the equal provider. Yes, it's going to probable be less difficult to get everything from one company, however you could discover which you like one type of product better from one seller than some other, and vice versa. Once you start to find out some carriers, it's time to gain out to them.

Now at this element, you don't need to buy something in bulk due to the reality you don't recognize what the product will actually be like. You want to revel in it for your self earlier than stocking your vending machine with it. So you'll need to mention which you're beginning a splendor merchandising device business business enterprise and which you're searching for to get a few samples to attempt out.

Depending on the corporation, sometimes you'll must pay for the samples and from time to time you received't. Don't be upset if you have to pay for samples because it's without a doubt really well worth the price. These wholesalers are corporations too and they don't need to get ripped off from people who need loose merchandise for private use.

I endorse checking out out merchandise from at least 3 unique agencies, however preferably 5 if you can. This will give you some attitude due to the fact you'll be capable of check the goods towards every precise. Once you find out a provider which you like, I advocate retaining at the least a 2 months supply of products available.

This manner you'll be ready for any emergencies. If considered one of your shipments gets misplaced or not on time, you'll nonetheless have some backups to inventory your device with. You don't need to attend till the final minute for the ones sorts of things.

Also, don't feel the strain to discover the extraordinary provider right off the bat. It should possibly take you a while as a manner to find out the proper in form for your commercial business agency. At first you would in all likelihood come upon a dealer that's precise sufficient, however not great.

Go earlier and buy some substances from them and hold looking for a higher dealer that you could beautify to. A lot of humans generally generally tend to get stuck of their strategies and that they neglect that they'll be fluid and exchange things within the event that they want to!

Chapter 5: Llc, Licenses, Insurance And Contracts

Simply beginning a vending machine commercial Enterprise Company without setting the proper parameters in area isn't always a smart idea. There is a piece of work you want to do to make sure that your corporation is operating efficaciously.

LLC

The first element I want to speak about is your organization entity. In the USA there are four special types of company entities, as follows:

C Corporation

S Corporation

Limited Liability Company (LLC)

Sole Proprietorship

If you're uncertain of which entity type need to you pursue, it's great to talk with a expert, in conjunction with a tax accountant or business enterprise corporation criminal professional. What I can will let you recognize is that C business enterprise are maximum typically used for big agencies. You may additionally have goals of increasing your vending tool commercial employer during the u . S . A . And in that case, a C enterprise business enterprise may moreover want to make experience for you.

You may additionally start your organisation business enterprise as some thing aside from a C commercial enterprise enterprise after which transfer it over to a C corporation as quickly as it becomes a larger business corporation. In the start even though, you'll possibly want some thing greater smooth. The wonderful course to transport is going to be an LLC.

This is probably an awesome desire as it technique you and your business corporation will not be regarded due to the fact the identical issue, like it would be within the case of a sole proprietorship.

That might not look like that massive of a deal to you, however while you and your business business enterprise are considered because the same, this shows your business enterprise property and your private assets are taken into consideration due to the fact the equal trouble. This way in case your industrial company were to end up in a lawsuit, your private belongings is probably at danger.

An LLC will assist to offer you safety, however, starting up an LLC isn't enough. You should act like a commercial enterprise business organization and now not act which include you and your employer are one within the same. One instance of what I imply via this is starting up a separate financial organization account on your business agency.

This separates your business business company' finances out of your personal price range. If you combo the two, then it seems like you're working a sole proprietorship and not an LLC. This is a big deal inside the case of a lawsuit because of what's apprehend as piercing the enterprise veil.

Generally talking, positive, established business business enterprise entities collectively with a C business enterprise, S enterprise, or Limited Liability Company can also have asset protection. However, if it may be confirmed which you're running your business like a sole proprietorship, then that protection can be pierced and your private belongings can come to be honest undertaking.

Insurance

Getting coverage is each different step you'll want to take. Anything can rise up and it's terrific to be organized just in case. Most places will want to see evidence of insurance in advance than they permit you to located your device on their belongings. Typically, the simplest type of coverage you'll need is sizeable jail duty insurance and it shouldn't charge you that plenty in keeping with month each.

Licenses

The licenses you'll need will variety relying on what town your gadget is going to be in. You moreover need to reflect onconsideration on extra licenses whenever you need to region a similarly tool in every different country or metropolis. So even as I can't let you understand what specifically you'll want, a number of the ones you'll want to appearance out for are going to be a supplier's permit and a industrial employer license.

Some extra specific licenses can be preferred relying in your place. You can determine which licenses can be wished via doing a web search for "town of clean". For example, if you live in Pittsburg, you'll search town of Pittsburg to be taken to the town's right internet website online.

From there you'll be able to discover touch facts so that you can determine what you want or even apply for a enterprise license as properly.

Contracts

The very last issue that needs to be blanketed is a agreement. This is something you'll need to have in area because it will assist to protect you. In maximum times, the commercial enterprise enterprise in which you're storing your vending device could have a settlement in area.

If they don't, then it's a clever idea as a way to provide an agreement to make certain that every parties are on the equal internet page. It's exceptional to lease an legal professional for this to make sure everything is done correctly. If you don't have the money for this, then locating any other location so that you can provide a settlement for you is an alternative.

So what are some assets you'll need to search for at the equal time as studying through the settlement to make certain that you don't get blindsided in a while down the road potentially? The first thing is going to be the term that your system is within the corporation. You can observe a predetermined length of time, together with 6 or twelve months, and then preserve to resume from there.

You'll want to have this in location to ensure that a person else can't take your spot by providing to pay greater. Your system can be there for the duration determined upon for the agreed upon rate. Speaking of expenses, be aware of that as properly.

Make sure the numbers are the best ones which have been stated and agreed upon. If no percentage of earnings have been included, make sure that there's no citing of it inside the agreement. Having a noncompete is probably a few factor that acquired't be blanketed in the initial settlement but you'll need to ask to have it added.

This will state that the corporation isn't allowed to usher in a few other beauty related vending machines into the business enterprise so long as your gadget is with that commercial enterprise. If the employer you're speakme to isn't inclined to comply to this, then it's fine to appearance some vicinity else as that is a pink flag. A organization cares about incomes income and they don't care approximately along with in more opposition if it technique they might make more money.

This will handiest hurt your income and it's a few component that may be averted, so make sure a noncompete clause is delivered within the agreement! These are the main belongings you'll want to take note of inside the agreement. When it involves the period of the hire, I recommend sticking to a few aspect shorter, collectively with 6 or three hundred and sixty five days.

Sure, your rate can also need to boom with the aid of the usage of a bit after this time period ends, however a shorter time period will offer you with more flexibility simply in case the region you selected subsequently finally finally ends up no longer being profitable. You'll then be capable of extra with out troubles pivot to a brand new place in region of being caught. If the area is worthwhile, then you could renew and appearance to signal an extendedterm agreement.

Chapter 6: Understanding Your Target Demographics

Your goal demographics can dictate the superb spots on your merchandising machines. If you're promoting protein shakes and electricity bars, setting your device close to gyms or fitness golf equipment will likely display worthwhile. Suppose your gadgets are greater satisfactory for extra younger crowds, like snacks and sodas. In that case, schools or college campuses may be extra excellent. But it is no longer just about the age corporation. Consider way of life, art work behavior, or maybe the time humans are most likely to use the merchandising device. Conduct surveys or acquire data if possible. The more you understand your ability customers, the better you may serve them, and serving them nicely begins with being in which they may be.

Accessibility Matters

The time period 'excessive-internet web page visitors area' is regularly thrown spherical at the same time as discussing pleasant places, but what does it suggest? Essentially, you want your vending machine to be in an area in which human beings manifestly pass thru or congregate. Train stations, place of business constructing lobbies, and network centers are remarkable examples. However, not in reality any immoderate-site visitors location will do; it desires to be to be had. If it's too bulky for people to acquire your tool, you're dropping capability corporation. The system ought to be visible and indoors an inexpensive strolling distance from the primary drift of foot internet web page site visitors.

Safety Concerns

Safety is a multi-dimensional trouble. Firstly, you need to make certain the vicinity is steady for clients. A location liable to criminal interest can't exceptional deter functionality clients but furthermore positioned your funding at hazard because of vandalism or theft. Secondly, the place need to be steady for whoever will restock the machine and gather the earnings. Well-lit regions, ideally with a few shape of protection or surveillance, are normally better alternatives.

Evaluating Competition

While you is probably willing to assume that having no opposition is ideal, the presence of different vending machines isn't usually a crimson flag. Other vending machines can propose a showed name for for that form of carrier within the location. However, too many comparable services can saturate the marketplace, making it difficult for any unmarried tool to be substantially worthwhile. Consider the prevailing machines' product strains—are they much like yours or precise enough to validate your get admission to into the gap?

Utilities and Logistical Considerations

You may additionally neglect it inside the planning phase, but make sure the vicinity has the favored utilities like electricity. Not all vending machines are the identical; some can also need more power or maybe a water line. Understand the machine's requirements and make certain the area can accommodate them earlier than you're making your decision.

Long-term Viability

Emerging locations with developing foot website online traffic can be golden possibilities for a merchandising device commercial enterprise organisation. Areas undergoing residential or corporation improvement are usually suitable bets for a protracted-lasting profitable place. However, it's essential to preserve an eye fixed fixed on any upcoming adjustments that would negatively have an effect to your organization. Things like introduction artwork blocking off get admission to to your gadget, or the ultimate of a prime enterprise business enterprise that drove foot site visitors, can dramatically have an impact in your profits.

In conclusion, finding the proper location is a good deal extra than clearly losing a pin on a immoderate-site visitors place on the map. It's a calculated choice that involves multiple variables, from expertise your target market to destiny-proofing your industrial corporation toward capability demanding situations. As you begin this challenge, don't forget: your industrial business enterprise isn't just merchandising machines; it's miles meeting a need, and to satisfy that want, you need to be wherein the people are. When you don't forget it that manner, you're now not simply selecting an area to location a device; you're selecting the degree on which your enterprise organisation will carry out. So, select accurately.

Strategies For Negotiating with Property Owners

Negotiating the terms of an area in your merchandising gadget includes an entire lot more than choosing a month-to-month rental price. It's a communicate in which you and the assets owner or supervisor aim to reap an settlement useful to each activities. Here's a way to navigate thru this vital step.

Firstly, whilst you method a assets owner, it's far crucial to come returned prepared. They're going to need to recognize what is in it for them—beyond simply the hire you're inclined to pay. They want warranty that having your vending system on their assets is a win-win scenario. That may additionally need to mean offering them a percentage of your profits, or likely imparting a few unfastened merchandise. It could also be that your vending device serves their clients or employees in a manner that gives fee to their established order.

Understand that for the assets owner, it isn't solely about cash; it is about the brought gain to their assets and likely their clients. Your merchandising gadget may additionally need to, for example, be a comfort for an place of business constructing, making it much less complicated for employees to capture a snack without leaving the premises. It may also be a small income purpose strain for a network center or perhaps a promoting thing for an condo complicated. Tailor your pitch to how the vending device meets those wishes, going past mere convenience to demonstrate its wider benefits.

Another negotiating thing is the term duration of your settlement. Short-term contracts deliver each you and the assets owner flexibility, however they'll moreover create instability to your business enterprise. Long-term contracts, inside the period in-between, offer a diploma of protection however may additionally additionally lock you into an destructive scenario if matters circulate south. Consider what you're comfortable committing to, however moreover be organized to conform to the alternatives of the belongings proprietor.

Then there can be the problem of utilities. Will you want to pay greater for the power your merchandising tool makes use of, or is that blanketed for your month-to-month lease or percent? Sometimes the ones are points you may negotiate, especially in case your merchandising device offers excessive-profits items like espresso or perishable goods that could trap more customers to the region.

Always keep in thoughts to place the whole lot in writing. A handshake settlement might not guard you if disagreements or misunderstandings arise down the road. The phrases need to be actually defined in a proper settlement, and reviewed by means of manner of a felony marketing and advertising representative acquainted with commercial leases or agreements. The settlement need to element the lease, application bills, income-sharing if any, phrases and conditions for termination, and distinctive crucial elements. This ensures transparency and lays the foundation for a expert dating among you and the property owner.

Finally, be organized for a few again-and-forth. Negotiation is a device, and it's miles uncommon for both occasions to agree on all phrases right away. Don't be too inflexible in your situations; go away some room for compromise. At the identical time, have your boundaries set for your thoughts. Know the lowest rent you may pay and the best percent you may deliver while although maintaining profitability.

In summary, negotiation with property proprietors is a nuanced approach requiring education, transparency, and the willingness to discover mutual blessings. When finished correctly, this step now not best secures a area on your merchandising tool however also establishes a foundation for an extended-lasting, worthwhile courting with the property owner. And in industrial organization, such relationships are definitely well worth their weight in gold.

Evaluating and Analyzing Location Performance

Once you have secured a place and your merchandising machines are operational, you can not sincerely sit down returned and desire for the quality. You need to always examine and examine how every vicinity is appearing. Think of it as tracking the important signs and symptoms of your commercial company; you need consistent tracking to make certain it stays healthful and grows. So, what metrics must you pay attention to, and how do you're making enjoy of them?

First, the exquisite and most direct metric is every day income. Every vending gadget is ready to track how many of every item is offered. These profits figures offer you with a photo of techniques nicely the tool is doing. But every day profits figures may be misleading. They can vary because of various factors like weather, vacations, or close by events, so it is essential to look at longer-term tendencies.

Besides the smooth rely variety of items offered, check what objects are selling. Are more wholesome snacks outperforming conventional ones? Are drinks promoting higher than meals devices? This form of records allows you optimize what you inventory your machines with, permitting you to satisfy call for greater efficiently and probable command higher income margins. You might be amazed to find out that an difficult to understand brand of tea outsells famous sodas at a particular area. Utilize this information to adjust your stock as a end result.

Location site site visitors is some other important metric. Even in case your vending device is absolutely stocked with famous devices, it is able to not depend plenty if people are not passing via it. To gauge foot internet site online visitors, you may use smooth strategies like counting how many human beings stroll beyond your device in every period. However, greater state-of-the-art strategies, like installing sensors or video analytics, can offer you with a clearer, greater extraordinary photograph. These era can also come up with insights into the demographics of people who not unusual the region, helping you tailor your offerings even greater exactly.

You ought to also pay attention to consumer remarks. Yes, merchandising machines are usually a 'silent company,' but that doesn't imply you can not get patron critiques. Placing a QR code at the system that results in a brief survey can yield worthwhile insights. You can ask questions like, "What specific products would you want to look?" or "How glad are you alongside facet your purchase?" Customer remarks can guide you on what to inventory or perhaps at the tool's usability. For example, if customers usually whinge approximately the machine failing to dispense merchandise, you realise you have got a mechanical trouble to deal with.

Operational fees for each area are also vital metrics. Each merchandising system region have to have one-of-a-type operational expenses based totally on factors like hire phrases, software expenses, and even the price to restock the system. Calculating the operational price regular with sale can give you a extra realistic photograph of ways profitable a specific place is.

Then there may be the stock turnover rate—the frequency at that you want to restock. This is a double-edged sword. High turnover is exquisite for cash flow however may want to probably boom your restocking prices. Low turnover, then again, approach your objects are sitting for a long time, which could reason expired merchandise or coins go with the flow issues. Finding a stability is important, and the most excellent turnover charge need to differ from one place to 3 different.

To make revel in of a whole lot of those metrics, you may probably want to spend money on merchandising tool manage software application application. This software application software can music sales, stock, and fantastic critical metrics, allowing you to look the entirety in a unmarried vicinity. With this records in hand, you may effortlessly have a look at the overall performance of numerous locations and make information-pushed choices to enhance your organisation.

In precis, tracking and comparing a couple of aspects of every merchandising gadget place is vital for the prolonged-time period achievement of your merchandising gadget business business enterprise. By cautiously tracking profits inclinations, client options, and operational prices, amongst different metrics, you may perceive which locations are most profitable and which would probable require adjustments or maybe relocation. It's all part of taking a proactive, in choice to reactive, method for your merchandising system business business enterprise.

Addressing Common Objections from Property Owners

Navigating objections from belongings proprietors is an inevitable part of the vending tool enterprise enterprise. As you pitch your machines to severa locations, you can encounter wonderful issues and reservations. Knowing a way to deal with the ones can be the distinction among landing a worthwhile spot or lacking out on an possibility. Let's discover some not unusual objections and the techniques to counter them.

Firstly, you may come upon property proprietors who are skeptical approximately the cultured effect of vending machines. They worry that a vending system may probably detract from the surroundings or the seen enchantment of their set up order. In those instances, provide to expose them contemporary-day-day, glossy merchandising tool designs that could be a praise, now not an eyesore. Many present day merchandising machines include customizable skins or lighting fixtures that may fit the place's current decor. Some even offer touch-show show screen generation, improving the overall purchaser experience and reflecting well on the assets.

Another commonplace objection is spherical region. Property proprietors frequently have constrained floor area and are hesitant to allocate a chunk of it to a merchandising tool. Here, the secret's to demonstrate the inexperienced use of space that cutting-edge merchandising machines provide. You may also want to, for instance, gift compact machines that healthy into corners or narrow hallways. Also, highlight that merchandising machines can offer an additional supply of income for the assets owner thru profits sharing or a flat condo rate.

Safety and protection are also big worries. Property proprietors can be hesitant to usher in vending machines due to fears of robbery or vandalism. To counter this objection, you could spotlight the safety capabilities contemporary-day merchandising machines come prepared with. This consists of sturdy production, solid fee gateways, or even included surveillance cameras. Additionally, you can offer to take complete responsibility for any damages or safety issues that could upward push up, assuaging the assets owner's worries.

Perhaps the most vital objection you can face is concerning profitability. Many belongings owners will question the sales capability and wonder if the distance may be used extra profitably. Here, facts is your great pal. Be organized to percent profitability studies or achievement memories of comparable locations. This assist you to paint a sparkly picture of the financial benefits, persuading property proprietors to take a risk on your vending device industrial commercial enterprise organization.

Sometimes, the objection is probably from present tenants or stakeholders within the property. They may in all likelihood fear that a vending machine will devour into their income or generate clutter. If it's far the case, be prepared to behavior a greater brilliant presentation that is composed of these stakeholders, demonstrating how vending machines can complement present organizations in vicinity of compete with them.

The key to overcoming objections is being organized and doing all your homework. Know the distinct varieties of merchandising machines and their competencies, understand the worries belongings proprietors would likely have, and feature organized solutions for every objection. Reassure them that your vending machine enterprise isn't always handiest worthwhile but moreover trouble-unfastened and beneficial for their assets and their purchasers. By doing so, you not simplest growth your opportunities of securing a pinnacle region however additionally construct a robust foundation for an extended-term commercial enterprise dating.

Creating Win-Win Agreements

The key to any a hit business organization relationship is mutual benefit, and your dealings with belongings owners aren't any exception. A win-win agreement is greater than virtually splitting profits or agreeing on rent; it is about creating a symbiotic relationship wherein both activities are incentivized to ensure the economic employer's fulfillment. How are you able to craft such an agreement in the merchandising gadget enterprise? Let's discover.

For starters, flexibility is critical. Property proprietors may additionally moreover have numerous dreams, regulations, or options, and being inflexible to your industrial organisation version won't serve you properly. Are they extra comfortable with a income-sharing version, or do they choose a flat month-to-month rent for the tool's region? Maybe they need to test the waters with a quick-time period settlement in advance than committing to prolonged-term. The greater adaptable your provide, the much less complicated it'll be to close the deal.

While we're about sales-sharing, it's crucial to be apparent about how profits are calculated and cut up. Whether it's far a set percent or a tiered gadget primarily based on income quantity, ensure both events are clean at the data. A properly-described, obvious association minimizes capability disputes down the line and guarantees a smoother relationship.

Do not underestimate the power of greater benefits. What else are you able to offer to make your concept greater enticing? For instance, you can provide to control restocking and renovation genuinely, releasing the property owner from any operational hassles. Alternatively, you could offer an initial "release merchandising" to draw extra customers to the region, which would possibly gain distinctive corporations on the property as properly.

To assuage concerns approximately the cultured or useful effect of the device, you can suggest an ordeal period. During this time, the assets proprietor can decide the device's effect on consumer web page visitors, ordinary earnings, and surroundings. If the trial is a fulfillment, it'll be much less complex to barter an prolonged-term settlement.

Legal factors need to additionally be meticulously addressed. The agreement have to spell out terms certainly, together with exit clauses, prison obligation issues, and dispute resolution mechanisms. Both parties ought to assessment and agree on the ones phrases, probable with the help of criminal advise, to ensure that they will be straightforward and obvious.

Lastly, continuously keep the lines of conversation open. A a success merchandising gadget industrial employer is based totally no longer just on client delight however moreover on preserving a healthy dating with assets proprietors. Regular test-ins, usual overall performance updates, and a willingness to pay attention and adapt can move an extended way in ensuring that the settlement continues to benefit each activities.

Creating a win-win agreement takes more than absolutely organisation acumen; it takes empathy, adaptability, and a eager records of your accomplice's goals and expectancies. When every sports feel like they may be gaining some component valuable, the path to a long-time period, worthwhile relationship becomes masses clearer.

Chapter 7: Purchasing And Managing Vending Machines

Guide to Purchasing Vs Leasing Vending Machines

One of the most pivotal selections you will make for your vending gadget mission is whether or not or now not to buy or rent your merchandising machines. This preference will set the trajectory to your operational expenses, your income model, or even your everyday control style. So, permit's get into the nitty-gritty of every alternatives that will help you make an knowledgeable desire.

Purchasing a vending tool may possibly appear to be a massive initial investment, however it comes with a degree of control and freedom that leasing could not provide. When you very very own the tool, you have got whole autonomy over what merchandise you inventory, the way you keep the system, or maybe in which you decide to region it. These factors can extensively have an effect for your earnings margins. Moreover, proudly proudly owning the device outright gets rid of monthly hire payments, permitting you to obtain profitability extra speedy. But it's crucial to don't forget that this option additionally manner you go through all the expenses and responsibilities for protection, restore, and enhancements.

Leasing, but, offers a manner to enter the marketplace with plenty tons less financial danger prematurely. Monthly lease bills can be extra possible for inexperienced human beings who are though studying the ropes of the agency. Additionally, some leasing contracts encompass upkeep and restore services blanketed, relieving you of these obligations. But the drawback is that leasing regularly comes with guidelines—you'll be restricted in the products you may inventory or perhaps in the locations wherein you could area the system. Plus, the ones monthly bills maintain for the lifestyles of the lease, doubtlessly ingesting into extended-time period profits.

Now, how do you decide which route is proper for you? Start with a radical monetary evaluation. Consider the whole charge of possession for purchasing a system—this consists of now not truly the tool's fee however moreover maintenance, upkeep, and any potential enhancements you could need to keep the machine aggressive. Compare this to the complete projected value of leasing, remembering to element in the lengthy-term nature of those monthly payments.

Risk tolerance is any other essential aspect. Purchasing can also provide higher returns ultimately, however it's also a more top notch initial funding, making it riskier, especially for the ones new to the commercial enterprise. Leasing is probably a more secure wager if you're although finding out the waters or in case you want to spread your startup capital during severa components of your organization, like advertising and advertising and advertising or more places.

You'll furthermore need to do not forget your prolonged-term enterprise technique. Do you advise to stick with one or machines, or are you making plans on scaling up all at once? If it's miles the latter, leasing could possibly offer more flexibility to swap out machines to fulfill market name for or take benefit of technological enhancements. Ownership makes extra enjoy if you're looking for a more static, extended-time period setup.

Another aspect to endure in mind is your knowledge set. If you are available and experience tinkering, preserving your machines can be cost-effective and could lean the scales inside the route of purchasing. If the very idea of a wrench makes you wince, a leasing settlement with renovation included might be your amazing wager.

Don't forget to take into account tax implications as well. Purchased machines can be depreciated through the years, offering a few tax blessings. In assessment, hire bills are normally virtually tax-deductible as a commercial company fee, but you could not gather any fairness within the system.

In summary, there may be no one-length-suits-all strategy to the query of purchasing instead of leasing. It comes proper all the manner right down to your financial state of affairs, your capacity set, your risk tolerance, and your extended-term commercial enterprise desires. Make first-rate to don't forget those sorts of elements carefully and talk with monetary and legal advisors to make the most knowledgeable preference.

Maintenance, Repairs, and Technological Upgrades

One difficulty is for fine: vending machines aren't a 'set it and overlook approximately about it' form of business enterprise. Proper upkeep, nicely timed protection, and maintaining up with technological upgrades are important for maximizing your profits and lengthening the lifespan of your machines. Let's unpack every of those vital regions that will help you streamline your operations and hold clients satisfied.

Firstly, regular preservation is not a luxury; it is a want. Dust and dirt can clog the tool's inner additives, primary to overheating and malfunction. A device that is out of order isn't always clearly an eyesore; it's far a hollow on your pocket. Schedule ordinary cleanings and inspections, and motive them to thorough. Check totally free wires, tired factors, and symptoms of wear and tear and tear and tear. If you're just beginning, you can no longer have the expertise to gauge what dreams checking. In this example, making an investment in professional servicing may be useful, as a minimum until you get the preserve of factors.

But however impeccable protection, topics can go awry. This is in which having a stable repair technique comes into play. Machine downtime is anathema to profitability in this corporation. The longer your machine is out of order, the extra income you lose. That's why it is important to have a listing of dependable repair services or perhaps an in-house restore individual if you're running on a bigger scale. When scouting for repair services, look for specialists familiar collectively along with your tool's make and model, and generally maintain a package deal of critical possibility components. This foresight can substantially reduce downtime and get your machine up and taking walks quicker.

And then there may be the difficulty of technological upgrades. Vending machines have come an extended way from mere coin-operated dispensers. Newer models characteristic touch monitors, mobile price alternatives, or maybe a ways off stock monitoring. As customers develop aware about those conveniences, vintage machines can seem plenty less appealing, probably affecting your sales. But updating might now not commonly mean looking for a present day day machine. Many machines may be retrofitted with more latest technology at a fragment of the fee of a brand new buy. Look into options for upgrading the price system to accept digital bills, or such as software software that lets in you to expose inventory ranges remotely. These enhancements no longer first-rate make your gadget greater attractive to clients however can also streamline your operations and stock management.

Having a technique for those enhancements is surely as essential as having one for renovation and renovation. Keep an eye fixed on organisation dispositions and customer selections. Annual exchange suggests and vending system expos can offer insights into upcoming generation and consumer behaviors. Networking with special marketers can offer precious sensible advice on what improvements are nicely really worth the investment.

There's moreover a cyclical element to enhancements. Let's say you have got delivered a today's charge device it genuinely is a achievement together along side your clients. That might also want to bring about improved income, which then justifies further investment in even more extremely-present day generation. It creates a virtuous cycle in which your commercial enterprise continuously evolves to meet and exceed client expectations.

Let's no longer overlook that every improve is a commercial enterprise business enterprise selection, this means that it need to make economic feel. Calculate the ROI of every improve by using manner of factoring in the improved sales you assume and compare it inside the direction of the charge of the enhance itself. Always hold your client base in thoughts. An decorate it is all the rage in business enterprise place of work houses may not generate the equal enthusiasm in a college campus setting.

Finally, document the entirety. Every protection test, each repair, and every enhance need to be meticulously recorded. This data is gold. Over time, it's going to help you see trends, expect troubles in advance than they turn out to be troubles, and make informed selections about future enhancements and investments.

In sum, a proactive technique to safety, safety, and technological upgrades can spell the distinction amongst a thriving merchandising device corporation and one which slowly bleeds coins. As the proprietor, it's miles your responsibility to make certain that your machines are typically in height state of affairs, no longer simply automatically, however moreover in terms of the person experience they provide.

Inventory Management and Selecting Products

Inventory control is type of a recreation of chess; every circulate you are making has a ripple impact at some point of the board. It's about balancing deliver and call for, optimizing stocking techniques, and choosing the proper merchandise to your intention market. Understanding this difficult dance can separate a a hit merchandising tool company from one that constantly struggles with overstocking or, worse, empty cabinets.

Starting with the basics: What exactly is inventory control in the context of vending machines? Essentially, it involves monitoring the products you have got in inventory, their expiration dates, and earnings charges. Unlike a traditional retail save, you can not come up with the cash for the expensive of extensive storage. Your "warehouse" is the device itself. Thus, you want to be specific in your calculations to make certain that you neither run out of well-known gadgets nor have slow-shifting merchandise occupying valuable real property interior your system.

A crucial trouble of inventory manipulate is understanding your clients. This isn't always handiest a rely of getting a well-known idea of what humans might possibly like. It entails studying looking for conduct, strolling surveys, and probable even doing A/B tests to determine out what merchandise resonate maximum at the aspect of your goal market. For example, a system positioned in a fitness center will possibly do properly with protein bars and electrolyte beverages, at the same time as one in a university constructing would possibly find out a organized marketplace for quick snacks and caffeinated drinks.

Selecting products is not quite plenty purchaser desire, although. You moreover want to preserve in mind elements like shelf lifestyles, stocking situations, and company reliability. For example, a warm company with a quick shelf existence can grow to be a felony obligation if not controlled cautiously. You can also need to reduce the price appreciably because of the fact the expiration date nears, affecting your income margins. On the possibility hand, devices with an extended shelf existence won't be as touchy to such issues however may additionally have their traumatic conditions, like decreased client interest over the years.

Cost is a few different large aspect in choosing merchandise. If a product costs you too much, your vending device charges can also come to be uncompetitive. Also, immoderate-price gadgets would possibly take a seat down longer in the gadget, tying up capital that would be used for unique, quicker-shifting merchandise. Therefore, an in depth price-gain assessment have to be your cornerstone method even as selecting products.

Given the ever-evolving client tastes and market trends, staying static isn't an desire. You'll want to rotate your product services often. Use the facts out of your profits and stock monitoring to section out underperformers and introduce new gadgets. You can also use seasonal topics; for example, hot cocoa in wintry climate and lemonade in summer time can be worthwhile short additions to your inventory. Always preserve an eye consistent on your opposition too. If they're selling a product that is continuously presented out, it is probably surely truly worth considering for your line-up.

Moreover, managing your stock isn't best a one-time setup; it's miles an ongoing system. Technology can be your great buddy here. Several merchandising device manipulate software program software program options to be had on the market can help you tune income, show inventory stages, and even supply you signs whilst it is time to restock. Using this type of tool can reduce guide errors, shop time, and give you precious insights into your business' performance.

Also, do no longer underestimate the rate of constructing sturdy relationships with vendors. Negotiate terms that gain every activities. For example, a few vendors also can additionally offer discounts for bulk purchases, whilst others might also provide greater flexible rate terms. A strong supplier dating also can give you a better negotiating characteristic in terms of issues like shipping instances or handling deliver chain disruptions.

So why is all this critical? Because stock management can both make or smash your merchandising gadget industrial corporation. Too heaps stock ties up your capital and will growth the hazard of merchandise going unsold or expiring. Too little inventory approach out of location profits possibilities and pissed off customers. Both scenarios will consume into your earnings and can probably jeopardize the sustainability of your business organization.

In precis, inventory control and product preference require a statistics-driven, systematic approach balanced with a keen understanding of your aim marketplace. These practices will not fine assist you maintain a clean operation however additionally assemble a loyal purchaser base eager on returning for your machines time and again.

Optimizing Sales and Marketing Strategies for Your Machines

When you step into the vending device commercial corporation, you are now not truely becoming an operator; you're basically entering into the shoes of a marketer. The system is your billboard, the product selection is your marketing advertising marketing campaign, and the purchaser enjoy is your logo. It's a microcosm that captures many elements of advertising and marketing, and if performed proper, it could offer you wealthy rewards.

Let's communicate about optimizing earnings first. One of the preliminary strategies you should attention on is pricing. The fee trouble can carefully have an impact at the looking for decision, however it's also closely linked in your industrial business enterprise expenses and favored income margins. While it is probably tempting to set excessive expenses to maximise income, doing so without providing perceived charge can alienate potential customers. Your pricing technique should, consequently, be a properly-balanced equation that factors in expenses, competition, and customer willingness to pay.

However, pricing is just the forestall of the iceberg. How you gift your merchandise within the gadget can also make a big distinction. For example, putting amazing-promoting or excessive-margin items at eye degree can growth their visibility and likelihood of buy. The equal common sense applies to bundling—presenting complementary items together at a barely reduced rate can encourage big transactions.

Technology, over again, involves the rescue whilst optimizing sales. Modern vending machines can offer touchscreen interfaces, cashless bills, or even AI-pushed tips based totally totally on preceding purchases. These technological enhancements now not pleasant decorate client experience however moreover acquire valuable statistics on buying conduct, top income times, and product normal average overall performance. For example, in case you examine that a particular product sells out interior hours after restocking, it might be absolutely definitely really worth dedicating greater area to it or maybe negotiating a better cope with the supplier.

Now, allow's transfer gears to marketing and advertising strategies. Your merchandising machine commercial enterprise organization might be small, however that does not recommend you cannot anticipate huge in phrases of marketing. One of the maximum effective gear you have got is the device itself. Customizing the device's outside to showcase your emblem's emblem, sunglasses, and taglines have to make it stand out. But don't forget, it should now not be overwhelming to the point that it discourages interplay; the reason is to draw, now not repel.

Beyond the physical elements, digital advertising and advertising can also offer a strong platform to increase your vending system industrial company's visibility. Utilize social media to announce new places and specific promotions, or maybe run purchaser surveys. Email newsletters are every other beneficial tool to preserve your consumer base knowledgeable and engaged. You can encompass updates, offers, or perhaps a laugh data about vending device facts or technology.

Partnerships are every other road well in reality worth exploring. Teaming up with the belongings owners wherein your merchandising machines are placed to run joint promotions can help beautify earnings. For example, a health club need to offer discounted power drinks out of your gadget as a part of a membership package deal deal. You also can collaborate with nearby corporations to offer unique gives, thereby increasing your device's foot web page site visitors.

Promotions and discounts may be specifically powerful but ought to be used judiciously. Running promotions too frequently can devalue your merchandise, even as sparse promotions won't generate enough interest. Seasonal or tour-primarily based absolutely promotions are regularly well-received. For instance, you could provide a 'Back to School' marketing supplying wholesome snacks and beverages at discounted fees. Such campaigns no longer remarkable energy earnings but moreover make contributions to brand building.

However, all the income and marketing optimization in the worldwide can fall flat if you do no longer measure overall performance and regulate because of this. Track the ROI of your severa advertising and marketing and advertising and marketing efforts to apprehend what is operating and what isn't. This may additionally require a few funding into analytics tools or structures, but the insights you advantage can be treasured for long-term achievement.

In essence, optimizing sales and advertising strategies in your vending system business agency is an ongoing system that goals constant hobby, creativity, and analytical wondering. As your organization grows, you will discover more opportunities and channels to increase your reach and sales. But bear in mind, the vital mind stay the equal: recognize your purchaser, provide charge and continuously try and enhance.

Chapter 8: Legal And Financial Aspects

Legal Considerations and Insurance

When taking walks a vending system commercial agency, crossing the T's, and dotting the I's legally cannot be an afterthought—it ought to be a priority. Operating in whole compliance with the law no longer best prevents costly jail troubles but also builds your recognition as a responsible enterprise enterprise proprietor. The first step in crook compliance is understanding the sorts of permits and licenses you can want.

In many jurisdictions, you will be required to gain a stylish enterprise license to characteristic. This is the foundational document that legitimates your enterprise enterprise. Depending on the locality and the character of your merchandising machine commercial corporation—which include the varieties of merchandise you're selling—you can moreover need additional allows. For example, if you're selling food or drinks, a fitness department permit is type of generally compulsory. The permits wanted can range notably from one jurisdiction to three exclusive, so it's far essential to looking for advice from nearby and kingdom government assets or even prison specialists to make certain you are truly compliant.

Once your allows are in order, take into account to resume them often. Many organization proprietors get caught off shield on the same time as their allows expire, main to operational hiccups that would resultseasily were prevented with a bit of foresight and making plans.

Insurance is another non-negotiable factor of taking walks a vending device business enterprise. At the very least, you could need famous prison duty insurance to cowl any damages or accidents that could get up concerning your merchandising machines. Property coverage can shield the machines themselves toward theft or damage, and when you have personnel, worker's compensation coverage will probable be a demand.

It's moreover realistic to preserve in thoughts employer interruption coverage. Vending machines are typically low-renovation, but they are no longer no-safety. Suppose considered one among your most worthwhile machines breaks down, or worse, numerous machines fail concurrently because of a software software software program glitch. In that case, the economic impact might be tremendous. Business interruption insurance can help mitigate these risks through manner of overlaying misplaced income at some point of downtime.

Documentation is essential in any business corporation however even more so on the identical time as there are crook implications. Keep organized records of all lets in, inspections, and insurance guidelines. Digital copies are handy, however having bodily backups saved in a steady place is a nice exercising.

Contracts and agreements with belongings owners wherein your vending machines are positioned want to moreover be meticulously drafted and saved. These need to define all responsibilities, which includes who is answerable for what within the event of robbery, vandalism, or notable problems. Clear contracts can save you future disputes and provide prison protection for each activities involved.

Also, undergo in mind information safety prison tips, specially if your merchandising machines are tech-savvy, and collect patron data for personalised marketing and advertising. Regulations similar to the General Data Protection Regulation (GDPR) in Europe can practice if you're gathering statistics from European residents, and non-compliance can bring about hefty fines.

In summary, searching after the crook and coverage factors of your merchandising device enterprise isn't just a chore you need to slog thru; it's an indispensable part of your advertising and marketing strategy that safeguards your investment and popularity. When dubious, attempting to find professional prison advice is an investment for your industrial enterprise's durability and success.

Financial Planning and Budgeting

One of the fundamental benefits of the vending tool organisation is its pretty low operational fee. Yet, low rate could now not mean no fee. Effective monetary making plans and budgeting are important for ensuring your challenge not most effective survives but flourishes. For many, the enchantment of the merchandising system enterprise is its obvious simplicity. Place a tool, stock it, and watch the cash roll in. While it is real that vending machines may be a honest way to generate profits, the simplicity frequently obscures the monetary intricacies that would make or damage your commercial enterprise. Let's smash down what financial making plans and budgeting entail for your vending mission.

First off, you will want startup capital. The initial investment will in large detail rely on the type of vending machines you're seeking to buy or rent, and the quantity of machines you must carry out. Will you be looking for new or used machines? Do you want specialised machines for perishable gadgets? These are essential questions that want to be answered to decide your initial funding. Additionally, you want to maintain in mind the fee of inventory, set up, and any permits or licenses you may want. All those elements must be thoroughly researched and documented for your advertising strategy.

Another tremendous in advance rate you can not forget about is the location charge. Many property owners will expect a percent of your income or a hard and fast month-to-month charge for the privilege of placing your vending gadget on their premises. Ensure you negotiate favorable terms, however preserve in thoughts that this may be an ongoing fee. Therefore, it's critical to forecast the anticipated income for each place to make sure the region charge may not eat your profits margins.

Now allow's communicate approximately operational fees. These consist of but aren't restrained to inventory restocking, system protection, and transportation. Fuel prices for trips to and out of your vending machine locations can upload up quicker than you might imagine. Maintenance must involve maintenance, updates, or maybe fashionable tool substitute within the case of irreparable damage. Ignoring or underestimating the ones ongoing costs can purpose unpleasant surprises and may drastically impact your profitability.

You'll moreover need to maintain in thoughts coins drift manipulate. Vending system companies frequently deal in small denominations, and it might be tempting to undergo in mind the coins within the machines as 'spending cash.' This is a volatile attitude. The money should be allocated for enterprise fees first and most important—collectively with restocking, system maintenance, and month-to-month costs. A separate organization account in your merchandising device project is not certainly an offer; it is essential for effective economic manage.

Once you have got calculated your expenses, it is time to don't forget your pricing method. You'll want to discover the sweet spot in that you price excessive sufficient to cover prices and make a income, but low sufficient that clients despite the fact that feel they'll be getting a good deal. Keep in thoughts that pricing could in all likelihood need to be adjusted relying on the location and motive demographic.

Budgeting is not a one-time project however an ongoing approach. You'll want to continuously evaluation your prices, profits sales, and internet earnings to make records-pushed picks. Implement accounting software to keep track of every greenback and cent this is going indoors and from your industrial employer. Use this information to update your advertising method and adjust your strategies due to this. Also, keep in mind seasonal inclinations—call for for first-rate merchandise might also moreover vary relying on the time of 12 months, and your budgeting have to account for this.

Last but now not least, in no manner underestimate the energy of an emergency fund. Machines damage down, theft takes place, and pandemics display up. An emergency fund acts as a economic cushion that could hold your merchandising tool company operational in hard instances. Financial experts frequently propose setting apart at the least 3 to 6 months' without a doubt well worth of running costs. While this might appear immoderate now, you will be grateful for this cushion if and while an emergency takes place.

To sum up, powerful economic planning and budgeting ought to make the distinction between a short-lived project and a protracted-lasting, worthwhile business agency. These are not components of your corporation to set and neglect; they may be dynamic factors that require ongoing interest and adjustment. By treating them as such, you location your merchandising machine enterprise up for sustained success.

Tax Considerations and Implications

Navigating the maze of taxes is a daunting venture for any corporation owner, and the vending machine business enterprise isn't always any exception. You may think that because you're dealing in small transactions—snacks, liquids, or possibly toys—the tax obligations won't be that complicated. Unfortunately, it clearly is a ways from the fact. As a vending system business proprietor, you'll want to recognize the community, kingdom, and federal tax implications of your project to make certain you are in complete compliance with the law. Let's find out the extremely good tax additives you ought to be privy to.

One of the primary tax-related troubles you may stumble upon is income tax. As you are promoting a product, you are obligated to accumulate sales tax on the ones income, and the fee will range depending for your jurisdiction. However, some states may additionally provide exemptions for sure styles of merchandise, like crucial meals gadgets. Being privy to the ones intricacies and preserving an updated file will be critical for proper tax reporting.

Then there may be the income tax. Both your us of a and the federal authorities will need to understand about your vending tool earnings. And allow's not neglect close by taxes, which can embody a number of levies, from city business corporation licenses to property taxes in case you very personal the distance wherein your merchandising machines are positioned. Your income after costs, which consist of the whole lot from restocking your machines in your adventure costs, may be undertaking to the ones taxes. This is why accurate file-retaining is essential.

Let's communicate depreciation. Vending machines are considered property, and as such, they depreciate over the years. According to IRS recommendations, vending machines typically have a seven-twelve months depreciation time table. You'll be able to deduct part of your merchandising system's value each 365 days, that might provide large tax advantages. Keep in mind, but, that the tips spherical depreciation can get complicated, and it is generally sensible to consult a tax professional to make certain you are maximizing your deductions on the same time as staying in the bounds of the regulation.

Your choice of commercial enterprise organisation shape may have an impact on your tax responsibilities. If you operate as a sole owner, your enterprise income can be dealt with as your private need to pay self-employment tax. On the alternative hand, in case you've installation an LLC or a Corporation, there are specific devices of tax responsibilities and benefits. Understanding the tax implications of your preferred enterprise shape can prevent an entire lot of cash and capability complications in the long run.

As we waft inside the path of a greater digital international, on-line income, and virtual merchandising systems are gaining traction. If you make a decision to diversify and begin selling your products online or via an app, you may want to bear in thoughts greater tax pointers related to on-line earnings. These can get complex right away, as you may need to address now not honestly federal however probably multi-u . S . Tax responsibilities.

A proactive approach to tax planning is essential. The tax panorama is ever-changing, with criminal hints and tips often updated. Being caught unprepared ought to bring about fines, consequences, and a big quantity of unnecessary strain. Thus, it is probably a clever investment to are attempting to find advice from a tax consultant familiar with the vending gadget company. They can help you navigate tax jail recommendations, find out opportunities for deductions, and assist you put together for tax season correctly.

Given the complexity of tax worries, the use of accounting software program designed for small corporations may also need to make a international of distinction. This software software will not only assist you maintain track of your earnings and prices but can also combine with tax software application application to simplify your annual tax schooling. Just make sure to replace your facts frequently, preferably at once after restocking your vending machines or incurring a agency charge.

In stop, taxes are an inevitable part of doing commercial enterprise agency, but with cautious planning, they do no longer need to be a stumbling block. Staying informed and prepared will bypass an extended manner in ensuring that your vending device business enterprise isn't always simply worthwhile however moreover compliant with all tax responsibilities.

Risk Management and Mitigation Strategies

Running a vending tool organisation may moreover look like a low-hazard challenge in comparison to different sorts of businesses. After all, machines do no longer name in ill or prevent with out be aware, right? While it's miles true that some risks are minimized on this commercial commercial enterprise employer version, that doesn't suggest it's far virtually without its personal set of worrying conditions that require strategic planning for danger mitigation. A merchandising tool left unattended may be damaged into, a region settlement can be terminated with out caution, or a tool ought to malfunction and reason unhappy customers. That's why a calculated technique to risk manage is essential.

Security is one of the essential troubles. Vending machines are often hassle to theft and vandalism. While it is almost no longer feasible to prevent all such incidents, making an investment in incredible safety features like superior lock systems, surveillance cameras, or alarms can deter ability thieves. A specific rule of thumb is to keep your machines nicely-lit and in areas with excessive foot web website online traffic, as remoted machines are much more likely to be focused.

Liability is each distinct vicinity that requires hobby. What if a customer receives unwell after eating a snack out of your gadget, or a very excited baby gets their hand stuck on the equal time as attaining for a toy? Liability coverage can cowl such mishaps, imparting a financial cushion and peace of thoughts. The types and quantity of insurance you may want can vary based totally at the styles of products you're presenting, the places of your machines, and one-of-a-kind elements, so ensure to go to an insurance marketing representative familiar with the vending tool industrial employer.

Operational risks are omnipresent. Your merchandising machines may additionally need to damage down, or there can be a power failure affecting your device's refrigeration device, inflicting perishable gadgets to wreck. Regular renovation and timely protection are essential. Keeping spare parts reachable and retaining a listing of reliable technicians can bypass an extended way in minimizing operational downtime. Additionally, the usage of merchandising machines with far off tracking capabilities will permit you to tune their ordinary normal overall performance and troubleshoot problems earlier than they grow to be exceptional troubles.

Chapter 9: Marketing And Branding

Creating A Strong Brand for Your Vending Machine Business

Branding isn't great a buzzword; it's the heart beat of a employer. A sturdy brand can differentiate your vending system commercial enterprise in a crowded marketplace, create a long-lasting have an effect on inside the minds of clients, and set up remember and loyalty. But how can a company that revolves spherical merchandising machines—which, on the start look, could in all likelihood appear impersonal—create a memorable brand? Here's a comprehensive manual to constructing a sturdy brand presence for your vending machine commercial enterprise.

The Power of Perception

The first step in branding is understanding its power. A brand isn't simply a emblem or a catchy tagline; it's the notion human beings have after they don't forget your organization. It encompasses their reviews, expectancies, and feelings tied to your merchandising provider. Whether it is the advantage of use, the nice of merchandise, or the responsiveness of your customer service, each touchpoint shapes this belief.

Define Your Brand's Core Values

Before you can translate your brand into tangible elements like trademarks, colors, and designs, you need to define your middle values. Ask yourself: What does my vending device company stand for? Maybe it's sustainability, with green machines and packaging. Perhaps it's miles innovation, providing the cutting-edge day tech or uncommon merchandise. Your center values act because of the fact the compass guiding all your branding efforts.

Designing with Purpose

Your logo, tool layout, or maybe the format of merchandise within the vending system play a pivotal feature in branding. They're the primary things customers see. Consider sunglasses that evoke the feelings you want to companion together together with your logo. For example, blue regularly symbolizes bear in mind and stability, on the same time as green can represent health or eco-friendliness. The format need to be regular in the course of all your vending machines and considered one of a type touchpoints like your internet internet site on line or marketing materials.

Crafting a Unique Selling Proposition (USP)

In a sea of vending machines, why want to a person choose out yours? Your USP is that compelling cause. It can be a totally unique product providing, incredible price, or a combination of severa elements. But it's critical to articulate this in a concise and catchy manner.

Consistency is Key

One of the important tenets of branding is consistency. Ensure that your merchandising machines, net net page, advertising substances, or maybe your social media profiles resonate the identical emblem message. Inconsistencies can confuse customers and dilute your logo's impact.

Engaging with Your Audience

In contemporary-day-day digital age, having a robust on line presence can enlarge your emblem. Engage along side your clients on social media, gather feedback, run contests, or percent within the returned of-the-scenes content material material about sourcing merchandise or preserving machines. These interactions humanize your brand and foster network.

Feedback Loop

Your logo isn't static; it evolves with time. Regularly acquire feedback from clients. Are there merchandise they had love to peer?

How's their normal enjoy? This feedback can offer insights now not handiest for enhancing your services but also for refining your brand.

Conclusion

Creating a sturdy logo to your vending device business agency is a combination of approach, layout, and consistent attempt. It's approximately telling your tale in a way that resonates together with your goal marketplace and devices you apart. As you enlarge and adapt to marketplace changes, your emblem can be the anchor retaining you rooted on your center values on the identical time as sailing within the course of fulfillment.

Offline and Online Marketing Strategies

In the age of virtual connectivity, the fusion of traditional offline advertising and marketing with online strategies is important to maximize the obtain of any organization, together with the merchandising system project. Both avenues encompass their amazing blessings and may function powerful equipment in promoting your corporation to a big target market. This section delves into effective strategies for every nation-states to make sure your merchandising tool employer remains visible, relevant, and attractive to capability customers.

Offline Marketing Strategies

1. Location-Based Advertising:

Promotion starts offevolved offevolved with the place of your vending device. Choose locations that see a high footfall however moreover consider places that relate to the goods you are merchandising. For example, putting a fitness snack vending device outdoor a gym will be a wonderful approach.

2. Brochures and Flyers:

These are value-powerful techniques to spread the phrase about your vending services. They can be dispensed in local community facilities, grocery shops, or at once to homes within the area of your vending machines.

3. Local Sponsorships and Partnerships:

Consider sponsoring a community sports activities sports activities group or community occasion. This no longer exquisite gives visibility on your business organization however moreover strengthens your ties with the community, portraying your emblem as one that cares and is concerned.

4. Guerrilla Marketing:

This unconventional form of advertising and marketing should make a huge effect. For example, staging a small occasion or project round your merchandising tool, providing unfastened samples, or taking component with nearby influencers can draw a crowd and generate buzz.

Online Marketing Strategies

1. Social Media Engagement:

Create business business enterprise profiles on systems like Facebook, Instagram, and Twitter. Regularly publish updates, run promotions, and interact collectively along with your lovers. Remember, it's miles not pretty an entire lot promoting; it's about growing a network round your brand.

2. Google My Business:

Claiming your corporation on Google can assist locals with out issues find out your vending device places. It additionally opens up the opportunity for consumer critiques, which may be instrumental in attracting new customers.

3. Email Marketing:

Collecting e-mail addresses and sending out everyday newsletters with promotions, updates, or exciting content material cloth can hold your target market engaged and remind them of your offerings.

4. Online Advertising:

Platforms like Google Ads or Facebook Ads allow for centered advertising and marketing, making sure your advertisements are seen by way of way of a demographic it sincerely is maximum in all likelihood to be inquisitive about your products.

5. Search Engine Optimization (SEO):

Having a internet website online on your merchandising gadget enterprise corporation can be a pastime-changer. Implement are seeking for engine advertising wonderful practices to make certain functionality clients can find out you outcomes while looking for merchandising tool offerings.

In conclusion, blending offline and online strategies guarantees a properly-rounded marketing and advertising approach, catering to a miles broader intention marketplace. In the evolving panorama of commercial organization, it is important to live adaptable, show the effectiveness of your strategies, and regulate as needed. Marketing isn't static; it's miles a dynamic, ongoing way that requires interest and innovation.

Analyzing Marketing Performance and Optimization

Marketing is an ever-evolving beast, in particular inside the realm of the vending tool business enterprise. It's critical no longer handiest to strategize however additionally to research and terrific-tune your technique continuously. Without keen insights into your advertising and marketing and advertising usual overall performance, even the most well-intentioned techniques can fall flat. Let's dive deeper into the area of marketing analytics and recognize the manner to optimize your campaigns correctly.

The Rationale Behind Marketing Analytics

Before diving into the 'how,' permit's recognition on the 'why.' Why is advertising and advertising analytics pivotal?

Informed Decision Making: Data-driven insights make certain your options aren't based mostly on intestine feelings however solid records.

Resource Allocation: By information what is strolling and what isn't always, you may direct your property successfully.

Enhanced ROI: Analyzing lets in in terrific-tuning strategies, making sure a better go back for your investments.

Customer Insights: Data well-knownshows patron behavior, options, and ache points.

Metrics That Matter

1. Return on Investment (ROI):

A pivotal metric, ROI measures the fulfillment charge of a selected marketing initiative in financial terms.

Formula: ROI = (Net Profit from Campaign / Cost of Campaign) x one hundred%

2. Customer Acquisition Cost (CAC):

Essential to understand the rate concerned in gaining new customers thru your marketing strategies.

Formula: CAC = Total Marketing Expenditure / Number of New Customers Acquired

3. Customer Lifetime Value (CLV):

Estimates the entire profits from a unmarried customer over their engagement period together collectively along with your corporation.

Formula: CLV = Average Purchase Value X Average Purchase Frequency x Average Customer Lifespan

4. Conversion Rate:

For virtual campaigns, this metric gives a percent of visitors who carry out the preferred motion, which include developing a buy.

five. Engagement Metrics:

Keep track of interactions on social systems. High engagement shows your brand's robust resonance collectively together with your target audience.

Utilizing Analytics Tools

A myriad of tools are available to simplify the analytics manner:

Google Analytics: A pinnacle device to gauge internet site engagement, presenting insights into vacationer demographics, conduct, and further.

Social Media Insights: Native to systems like Facebook and Twitter, they provide a deep dive into put up standard overall performance and target market engagement.

Feedback Platforms: Tools like SurveyMonkey offer direct insights from clients.

Sales and CRM Tools: Software like HubSpot offers insights into income inclinations and lead conversions.

Steps for Effective Optimization

1. A/B Testing:

Also known as split trying out, it consists of evaluating two versions of a campaign to determine which one is more powerful.

2. Customer Segmentation:

Grouping your clients based totally on shared trends can bring about more personalized and impactful campaigns.

3. Retargeting Campaigns:

A digital technique to re-engage site visitors who showed hobby but did no longer convert.

4. Periodic Feedback:

Consistently are searching for comments from customers and stakeholders. Use this precious records to make non-stop enhancements.

Remember, the energy of advertising and advertising and marketing analytics and optimization lies of their non-stop software. As the marketplace evolves, so have to your strategies. By generally reading and optimizing, you make sure that your merchandising gadget commercial enterprise enterprise stays beforehand of the curve and reaps the rewards of knowledgeable preference-making.

Building Customer Loyalty and Retention Strategies

The authentic achievement of a merchandising system organisation isn't really attracting new customers however retaining them and turning them into unswerving advocates. A loyal client not most effective brings in consistent revenue but moreover allows in herbal advertising thru phrase of mouth. Let's delve into the intricacies of constructing patron loyalty and strong retention techniques.

The Significance of Loyalty in the Vending Machine Industry

For many, vending machines provide comfort. Yet, in a market saturated with options, why should a patron pick your system over the simplest right subsequent to it? The answer lies in client loyalty. A system that constantly presents amazing, reliability, and in all likelihood a piece wonder now and then, will continuously be a favourite.

Predictable Revenue Stream: A reliable purchaser base ensures steady earnings, cushioning your commercial enterprise business employer inside the route of market fluctuations.

Lower Marketing Costs: Acquiring a trendy patron can cost as plenty as five instances greater than maintaining an present one.

Organic Growth: Loyal customers regularly become brand ambassadors, bringing in new customers with out greater marketing expenses.

Understanding What Drives Loyalty

Before embarking on constructing loyalty, it's far crucial to recognize its pillars:

1. Quality: Ensure your merchandise are of top-notch remarkable. Whether it is a beverage, snack, or any other item, maintaining constant terrific is paramount.

2. Reliability: Machines must be often checked for malfunctions. There's not some

thing more demanding for a consumer than a tool that takes cash but doesn't dispense merchandise.

three. Engagement: Use era in your benefit. Modern merchandising machines can provide interactive reviews, custom designed gives, or loyalty factors.

Crafting a Robust Loyalty Program

One of the simplest strategies is introducing a loyalty software. Here's a step-by means of way of-step guide:

1. Set Clear Objectives: Before diving in, define what you choice to accumulate. Is it more sales, accelerated customer engagement, or statistics collection for customized advertising and marketing?

2. Simple Yet Rewarding: An overly complicated software software can deter participation. Make certain the earning and redemption techniques are sincere.

three. Multi-tiered Systems: Offer particular ranges to your loyalty application. As clients skip up stages, they release more rewards, which evokes repeat company.

4. Personalized Offers: Use client statistics to offer personalised offers. If a consumer often purchases a specific snack, provide a reduction on that.

5. Digital Integration: Modern merchandising machines can be integrated with apps. Allow customers to music their factors and redeem rewards thru a dedicated application.

Beyond Loyalty Programs: Building an Emotional Connection

A emblem that resonates emotionally has a tendency to have extra dependable clients. Here's a manner to installation that connection:

1. Storytelling: Share the tale inside the lower back of your vending device industrial

employer. Was it a circle of relatives project? Was there a unique task you overcame? A relatable tale could make customers feel more related.

2. Community Involvement: Sponsor neighborhood activities or have interaction in network services. Being recognized as a logo that gives lower lower back can earn you extensive goodwill.

3. Feedback Channels: Always have a channel open for comments. When customers feel heard, they feel valued.

Strategies for Effective Retention

Loyalty and retention move hand in hand. To ensure clients hold coming once more:

1. Monitor Interaction: With covered technology, tune client alternatives. Do they buy a selected drink every morning? Offer them a discount now and again.

2. Address Grievances: If a customer has an problem, address it immediately. An green

redressal system can turn a upset consumer right into a devoted one.

3. Periodic Surveys: Engage with customers through periodic surveys. It gives them a sense of involvement and affords them with worthwhile insights.

four. Quality Control: Regularly rotate and test the products on your vending gadget. Freshness is a key determinant of repeat purchases within the merchandising device industry. Even one instance of a stale product can deter a consumer for a long term.

five. Engage on Social Media: Even no matter the reality that merchandising machines are via and large offline groups, there's splendid rate in organising a digital presence. Share updates, new product launches, or any adjustments in gadget places. Run contests or polls asking customers what that that they had need to see in the merchandising gadget next.

6. Surprise and Delight: Occasionally offer surprise offers or products. Maybe sometimes, a patron gets a two-for-one deal or discovers a brand new product pattern with their buy. These little unexpected joys can enhance the patron experience manifold.

7. Clear Signage: Ensure that your merchandising machines are prepared with clean instructions at the manner to apply them, the pricing, and any ongoing gives or promotions. Confusion or ambiguity can shy away potential repeat clients.

Leveraging Technology for Customer Retention

Technology can play a pivotal function in improving patron loyalty:

1. Mobile Wallets & Payments: Integrate with well-known cellular wallets to offer seamless transactions. This moreover allows cashless transactions, which can be a prime comfort difficulty for lots.

2. QR Codes: Place QR codes on your vending machines which, on the equal time as scanned, offer loyalty factors or direct customers to your on-line systems.

three. Machine Maintenance Alerts: Use sensors and predictive analytics to decide at the same time as a device is probably nearing a malfunction. This proactiveness can prevent capacity customer dissatisfaction.

4. Chatbots & Support: If you have got were given a net website or an app, implement chatbots. These can straight away deal with not unusual queries or problems, developing purchaser pride.

Evaluating Loyalty & Retention Metrics

Finally, as with every commercial organisation method, it's important to assess the effectiveness of your loyalty and retention efforts:

1. Customer Churn Rate: Monitor the charge at which customers stop the usage of your vending machines. A immoderate churn charge would possibly probable suggest an underlying hassle.

2. Customer Lifetime Value (CLV): This metric will come up with an understanding of ways plenty a purchaser is surely properly really worth for your business organization over an prolonged length.

three. Net Promoter Score (NPS): Regularly gauge the probability of your clients recommending your merchandising machines to others. It's a direct measure of patron satisfaction and loyalty.

four. Redemption Rate: If you have have been given a loyalty software in vicinity, reveal how frequently rewards or elements are redeemed. Low redemption can recommend that this device isn't resonating nicely.

In quit, customer loyalty is not quite a lot repeat business enterprise however constructing a dating. It's about developing memorable reviews round your merchandising gadget commercial enterprise, making customers experience valued, and making sure they continuously have extraordinary institutions with your logo. A loyal patron base is a testomony on your employer's splendid and company, and with the strategies outlined above, you could domesticate and nurture this valuable asset.

Chapter 10: Scaling Your Business

Strategies for Expansion and Growth

The proper mark of a a hit merchandising machine business business enterprise isn't sincerely the way it starts offevolved, however the way it grows. Scaling, or the approach of growing your employer operations, is an formidable organization that requires a touchy balance among chance-taking and strategic making plans. The praise? Increased earnings, more market presence, and a emblem that resonates on a larger scale. Let's discover the techniques to gather this growth.

1. Understanding the Right Time to Scale

Stable Operations: Before thinking about expansion, make sure your modern-day operations are solid. Consistent profits, a devoted customer base, and easy operations are indicators that you're ready.

Market Demand: Analyze the market to decide if there may be an increased name for on your vending services. Use surveys, remarks, and profits records to gauge this.

Financial Health: Ensure you have got the capital required for expansion, be it for purchasing new machines, securing new locations, or hiring extra team of workers.

2. Geographic Expansion

Adjacent Locations: Begin through the use of thinking about locations adjoining to your gift ones. This permits for much less tough control and oversight.

Research New Areas: Understand the demographics, foot web site site visitors, and client conduct of new areas in advance than setting your machines there.

Local Partnerships: Collaborate with community groups or companies within the new region to benefit a higher know-how of the marketplace and to make certain smoother operations.

three. Diversifying Product Offerings

Survey Existing Customers: Gather remarks on what extra products they'd need to peer to your merchandising machines.

Test New Products: Before a whole-scale roll-out, take a look at new merchandise in a pick out sort of machines to gauge their overall performance.

Stay Updated with Trends: The snack and beverage organisation is normally evolving. Keep an eye fixed on famous tendencies, be it fitness snacks, herbal merchandise, or present day drinks, and encompass them into your offerings.

four. Franchising Your Business

Standardize Operations: For a a success franchise model, you want to standardize every trouble of your business corporation, from tool placement to inventory control.

Legal Considerations: Work with prison experts to draft franchising agreements and recognize the regulatory elements.

Support & Training: Provide education programs for franchisees to make certain regular service awesome and logo example.

5. Technology & Automation

Invest in Modern Machines: Newer vending machines consist of advanced talents which embody virtual bills, inventory monitoring, and earnings analytics.

Remote Monitoring: Use generation to display screen inventory levels and income statistics in actual time, considering timely restocking and records-driven desire-making.

Automated Customer Service: Consider chatbots or computerized helplines to deal with customer queries, liberating up belongings for extraordinary duties.

Scaling a vending machine corporation is a exciting adventure that opens a worldwide of opportunities. However, with extra scale comes greater responsibility. It's essential to keep the identical degree of strength of mind, issuer nice, and customer interest as you enlarge. Every new tool, every new location, and every new product imparting must resonate with the center values that made your initial assignment a success. With careful planning, strategic investments, and an unwavering dedication to excellence, your merchandising device empire can acquire new heights.

Automating and Managing Staff

In the cutting-edge global of business enterprise corporation, automation, and powerful employees manage stand because of the reality the twin pillars that underpin the fulfillment and scalability of ventures, specially inside the vending device region. While automation can restrict guide responsibilities and optimize operations, a committed and nicely-managed group amplifies the performance and drives the commercial enterprise within the course of increase.

The Significance of Automation in Vending Machine Business

Efficiency and Precision: Automation, via its very nature, reduces the possibility of human errors. Whether it's approximately restocking merchandise, tracking profits, or reading overall performance metrics, automation ensures that operations run without troubles and continuously.

Cost Savings: While there may be an initial investment in integrating automation equipment or software program software, ultimately, automation can cause sizeable financial savings. With responsibilities being computerized, there's an lousy lot less need for manual intervention, which translates to decreased difficult work prices.

Data-Driven Insights: Modern vending machines equipped with virtual interfaces can provide valuable data approximately client picks, profits tendencies, and stock stages. This statistics, while processed thru automation gadget, can yield insights that can be used to refine business enterprise techniques.

Remote Management: Automation permits commercial organization owners to govern their merchandising machines remotely. Be it tracking stock ranges or profits, proprietors can get actual-time updates, making it less complicated to make knowledgeable decisions.

Staff Management: A Critical Element

Hiring the Right People: It's essential to lease frame of employees who apprehend the vision of your vending gadget business agency. Employees who resonate together along with your enterprise company ethos will not handiest be more powerful however will also contribute to growing a notable logo photograph.

Training and Development: Continuous training ensures that your employees is updated with the contemporary-day day in vending tool generation and customer support necessities. It's an funding which can bring about elevated profits and patron loyalty.

Creating a Positive Work Environment: A inspired and satisfied employee can be one of the most huge property for a merchandising tool business enterprise. Ensuring a stable, remarkable, and inclusive paintings environment can bring about increased employees retention and productiveness.

Streamlined Communication: Open channels of conversation amongst manage and frame of people can cause higher understanding and smoother operations. Regular meetings, feedback classes, and open-door guidelines can foster an surroundings of accept as proper with and collaboration.

Balancing Automation with Human Touch

While automation can result in standard overall performance, it's far vital to recognize that the human detail can not be in reality changed. For example, at the same time as a tool can restock products, it's miles the staff that guarantees the goods are glowing, the machines are smooth, and customers are served with a grin.

Moreover, in situations wherein a purchaser has comments or a grievance, a human touch – information and empathy – ought to make all of the difference. It's approximately hanging the proper stability – leveraging technology for normal performance whilst making sure that the human connection remains intact.

Conclusion

Scaling a merchandising device business business enterprise entails a symbiotic relationship among automation and human intervention. While technology can streamline operations, it's miles the human beings in the back of the machines that clearly pressure a commercial corporation in advance. By making an investment in each those factors, merchandising device marketers can make certain that their enterprise not pleasant grows but moreover stands the check of time.

Purchasing Existing Vending Machine Businesses

Entering the arena of merchandising machines doesn't always require beginning from scratch. Many entrepreneurs locate charge in purchasing modern merchandising device groups, leveraging set up routes, consumer bases, and operational structures. This technique can expedite profitability, but it is crucial to approach this road with a whole knowledge and a keen eye for due diligence.

Benefits of Purchasing an Established Business

1. Immediate Cash Flow: Unlike starting a latest industrial enterprise in which you'll in all likelihood must appearance in advance to months or perhaps years to appearance a earnings, shopping for an cutting-edge-day enterprise frequently technique you'll have exquisite cash go along with the go with the flow from day one.

2. Established Customer Base: An modern-day vending device course may additionally have a fixed of dependable clients. This set up purchaser base can provide constant income, decreasing the chance related to new ventures.

three. Existing Business Operations: From provider contracts to maintenance schedules, a longtime business enterprise may additionally have techniques in area, allowing you to consciousness on growth in desire to everyday operations.

4. Trained Employees: If the company has personnel, they may already be acquainted

with the operations, decreasing the time and belongings you'll spend on education.

Conducting Due Diligence

It's essential to make sure that the organisation you're thinking about is every legitimate and a valid investment. This includes an intensive exam of all components of the commercial organization:

1. Financial Records: Review financial statements, tax returns, and profits records for at least the beyond three years. This will provide you with a clean photo of the commercial corporation's profitability and any capability financial worries.

2. Legal Considerations: Ensure that each one licenses and lets in are contemporary-day. Check for any pending or capability prison disputes. This can encompass issues with landlords, providers, or clients.

3. Machine Condition: Inspect the physical scenario of the vending machines. Older machines might probable require extra safety or might not be as green as extra latest models.

4. Inventory and Suppliers: Check the prevailing stock and its age. Review contracts with vendors, making sure that the terms are favorable and there is no coming near near growth in expenses.

five. Reputation: Talk to customers, carriers, and landlords to gauge the popularity of the commercial employer. Online opinions can also offer insights into purchaser delight and potential regions of state of affairs.

Negotiating the Purchase

Once you have got determined that the industrial organisation is a incredible in shape, it's time to negotiate the acquisition. Factors to don't forget encompass:

1. Price: Based in your due diligence, determine a honest rate. If there are additives of the commercial enterprise employer that problem you, those can be factors of negotiation.

2. Financing: Discuss phrases of rate. Will or no longer it is a lump-sum price, or will the vendor provide financing? Ensure that any financing terms are easy and honest.

3. Transition Period: It is probably useful to have the preceding proprietor live on for a fast period after the sale. This can ensure a smoother transition, especially in phrases of client relationships and operational nuances.

Potential Challenges and How to Address Them

1. Cultural Shift: If the industrial company has personnel, they is probably used to a particular manner of doing topics. It's crucial to talk any adjustments correctly and make certain a smooth transition.

2. Operational Adjustments: While there is probably cutting-edge operations in area, there's continuously room for improvement. Regularly compare strategies for performance and profitability.

3. Customer Retention: Ensure that the transition doesn't disrupt the issuer clients are used to. Maintain communique and make sure that they are privy to any adjustments, emphasizing improvements.

Conclusion

Purchasing an current merchandising tool commercial business enterprise may be a worthwhile project, however it calls for meticulous hobby to element. By undertaking thorough due diligence, knowledge the blessings, and being aware of capability traumatic situations, entrepreneurs can characteristic themselves for success within the merchandising device organization.

Diversifying Your Product Offerings

In the dynamic global of vending machines, staying relevant and competitive necessitates an evolution in product services. Diversity in products might no longer pleasant cater to a far broader patron base however additionally mitigates dangers related to relying on a completely unique product line. Here, we'll delve into the intent, techniques, and actionable steps for product diversification.

Why Diversify?

1. Expanding Customer Base: By offering numerous merchandise, you faucet into super market segments and consumer options, ensuring your vending device appeals to a broader goal market.

2. Seasonal Variations: Some products would possibly sell better at some point of high-quality instances of the year. For example, bloodless beverages may also see a surge at some point of summers, even as

warmness drinks or soups can be in name for at some point of winters.

three. Risk Mitigation: If one product type sees a dip in profits, having a various range can compensate for that downturn.

four. Leveraging Trends: The marketplace is ever-evolving. Today's health-aware clients may additionally additionally select organic snacks, protein bars, or even glowing fruit over conventional chips and candies.

Steps to Diversify Effectively

1. Market Research: This is paramount. Before introducing new products, conduct surveys or remarks training. Understand what your customers need and what they revel in is lacking.

2. Test and Iterate: Instead of overhauling your entire product lineup, introduce one or new merchandise and gauge the reaction. Based on income data and feedback, you can then make more informed decisions.

three. Vendor Partnerships: Collaborate with diverse carriers. This not excellent gives you get admission to to a plethora of products however furthermore permits you to barter higher phrases and reductions because of bulk purchases.

four. Technology Utilization: Modern vending machines come geared up with digital interfaces. Use this to sell new products, provide combination offers, or provide discounts on less famous gadgets to move inventory.

Challenges in Diversification

1. Stock Management: Introducing a large kind of products could make stock control tough. Ensure you've got got were given a strong system in vicinity to song earnings, restock correctly, and avoid product wastage.

2. Product Shelf Life: Different products have incredible shelf lives. Fresh products like end quit end result or sandwiches want

not unusual replenishing. It's critical to recall this at the same time as diversifying to avoid wastage and make sure freshness.

3. Customer Overwhelm: While range is proper, an excessive amount of preference can every so often weigh down customers. It's a pleasant balance among supplying range and ensuring the selection device stays honest for the consumer.

four. Space Constraints: Vending machines have restrained area. When diversifying, bear in mind the size of the products and the way they healthy into your machine. Some merchandise would possibly require extra area, affecting the number of objects you may stock.

Tips for Success

1. Stay Updated: Markets exchange, and so do consumer options. Regularly evaluation your product services to make sure you're in track with modern-day tendencies and needs.

2. Listen to Feedback: Customers are your exquisite critics. Be open to comments, every outstanding and horrible. It offers treasured insights into what is operating and what should probably need a alternate.

three. Flexibility is Key: While it's miles important to have a plan, be prepared to pivot if something isn't always operating. Being flexible and adaptable is essential inside the speedy-paced merchandising device organization.

In sum, diversifying product offerings is a strategic circulate that, even as finished right, can considerably bolster income and purchaser satisfaction. By records the why and the way of diversification, merchandising device entrepreneurs can ensure their machines stay attractive and profitable in severa market conditions.

Chapter 11: Overcoming Common Challenges

Addressing Common Vending Machine Problems

Every commercial employer faces disturbing situations, and the merchandising device organisation isn't any exception. While this task gives a plethora of benefits, from minimal overhead to flexibility, it additionally has its precise set of issues. But fear not! With a proactive technique and knowledgeable strategies, you may without difficulty navigate those hurdles. Let's delve deep into commonplace vending device problems and discover the answers that would keep your organization thriving.

1. Machine Malfunctions and Breakdowns

Perhaps the maximum apparent undertaking on this agency is the gadget itself. These are mechanical devices, and like all machines, they might damage down or malfunction.

Solution: Regular renovation is fundamental. Instead of watching for a trouble to arise, time table ordinary checks. This proactive approach can choose out issues earlier than they grow to be whole-blown troubles. Also, generally have a relied on technician on call. Quick response times can reduce tool downtime, ensuring non-prevent revenue.

2. Out-of-Stock Items

An empty slot in your vending device is a overlooked sales possibility. Plus, it leaves clients disenchanted.

Solution: Implement stock monitoring structures. Today, severa software program application options can provide you with a warning at the same time as stock is on foot low. Furthermore, statistics your customers' alternatives and purchasing styles will let you stock the proper merchandise within the proper quantities.

three. Vandalism and Theft

It's an unfortunate fact, but merchandising machines can once in a while be goals for vandalism or theft.

Solution: Choose your locations wisely. Machines positioned in properly-lit, immoderate-website traffic areas are lots much much less probable to be centered. Also, invest in safety features like surveillance cameras or reinforced casing.

4. Currency and Payment Issues

From not accepting banknotes to rejecting cash or electronic bills, the ones issues can turn capacity income away.

Solution: Regularly check all charge systems. Moreover, preserve in mind diversifying your fee techniques. With the upward thrust of digital wallets and contactless bills, having numerous price alternatives can improve earnings and decrease rate-related troubles.

5. Product Expiry

Food devices, especially, have a restricted shelf lifestyles. The closing issue you want is a purchaser getting an expired product from your gadget.

Solution: Regularly rotate your stock, ensuring that older products are on the the front. Additionally, use stock manipulate structures to tune expiration dates and eliminate objects earlier than they're not steady to eat.

6. Price Setting Challenges

Set the charge too excessive, and clients could likely walk away. Set it too low, and your profits undergo.

Solution: Conduct normal marketplace research. Understand what opposition are charging and what customers are inclined to pay. Regularly examine and regulate expenses based absolutely on your expenses, opposition, and consumer name for.

7. Difficulty in Securing Prime Locations

A vending device's success regularly hinges on its place. Yet, finding and securing those excessive spots may be difficult.

Solution: Build robust relationships with property managers and owners. Offering a percentage of income can be a win-win. It's furthermore nicely certainly worth considering tons much less traditional but excessive-website site visitors places, like network centers or huge administrative center complexes.

In the arena of vending, annoying situations are honestly opportunities in hide. Each problem has an answer waiting to be discovered. With patience, patience, and a touch ingenuity, you may turn those common vending tool problems into growth catalysts on your company. Remember, each hurdle triumph over now not most effective strengthens your current assignment however furthermore paves the manner for future growth and achievement.

Managing Customer Reviews and Feedback

In contemporary interconnected worldwide, purchaser remarks is greater than best a tool for development; it's miles a lifeline for any commercial corporation. For vending machine operators, this comments becomes a useful useful resource, losing slight on areas that require hobby and people which might be excelling. Understanding a way to manage and leverage the ones insights is essential.

Starting a merchandising device commercial enterprise may also seem honest. You pick out out an area, installation your device, stock it, and look beforehand to the earnings to roll in. However, on this apparently easy equation, one component can substantially have an impact on fulfillment: the purchaser. Their choices, behaviors, and, most significantly, their comments can make or damage your project.

Imagine this case: You've installation a device in an place of business building. Sales had been accurate, but in the end, a client leaves a assessment mentioning that the device frequently rejects cash or that a particular item has expired. Left unchecked, such reviews can deter potential customers, leading to a decline in profits. On the turn detail, excellent opinions can enhance your organisation, making it crucial to actively control this comments.

Engaging with client opinions does extra than genuinely address man or woman issues. It sends a message that you care, that you're attentive, and that you're dedicated to offering the amazing possible issuer. Even a smooth acknowledgment of a excellent overview can create a ripple effect, predominant to more glad customers and progressed word-of-mouth advertising.

However, dealing with feedback isn't always quite an awful lot responding to evaluations. It's approximately the use of them as a device for boom. Negative critiques, while disheartening, can be specially insightful. They pinpoint regions of improvement, guiding your efforts and assets to in which they will be most needed. For instance, regular remarks approximately a particular product being out of stock can prompt a examine of your restocking time desk or perhaps lead to the attention of a secondary supplier.

But it's miles now not pretty tons the reactive measures. Proactively looking for comments can provide insights that customers may not spontaneously percentage. Consider putting in a suggestion container next in your machines or going for walks occasional online surveys. These proactive steps can assist discover capability troubles earlier than they growth, permitting you to deal with them all of sudden.

Building a relationship at the side of your customers is paramount. Consider imparting reductions or freebies to folks who depart comments. This now not only encourages more reviews however additionally fosters goodwill. Over time, as you gather a repository of feedback and actively paintings on addressing the troubles raised, you could discover that your vending device commercial enterprise isn't always just developing in profits but additionally in consumer pride.

In give up, on the identical time as the vending device business enterprise is probably rooted in automation, the human touch remains crucial. In the age of straight away communique, coping with purchaser critiques and comments isn't always simply an delivered mission; it is a necessity. By valuing every piece of feedback, whether outstanding or horrible and with the resource of actively operating in the path of developing a higher client revel in, you solidify your business's basis, ensuring its sturdiness and success.

Handling Objections and Rejections

In the place of organization, objections and rejections are inevitable. Especially in the vending machine enterprise, in which securing top locations or introducing new merchandise can face resistance. While these setbacks can be disheartening, they frequently represent opportunities in cover. The key lies in expertise, addressing, and transforming them into boom drivers.

Every entrepreneur will face objections, be it from property owners who are reluctant to permit a merchandising gadget on their premises or from customers who would probable have a criticism about a product. It's now not the objection itself however the way you manipulate it a good way to outline your business employer's trajectory.

Let's delve deeper into the common objections you can come across and the manner to navigate them.

Objections from Property Owners: Convincing a assets proprietor or supervisor to will let you place a merchandising device may be tough. Common concerns include functionality disruptions, protection issues, or definitely aesthetics. Addressing those objections head-on is critical. Present them with statistics taking walks web website online site visitors and profits projections, showing them the capacity revenue they may earn from commissions. Highlight the minimal noise and disruption modern-day-day machines make and the normal protection schedules you may adhere to. By turning their objections into blessings, you can often win them over.

Consumer Objections: Consumers may item to product pricing, awesome, or perhaps the selection available. Regularly reviewing consumer remarks can assist count on the ones objections. For pricing troubles, recollect offering promotions or loyalty programs. If product satisfactory is in question, assessment your companies and make sure your inventory is easy. Always be open to product tips, demonstrating that you price patron enter.

Regulatory Objections: Sometimes, close by regulations would possibly probable pose challenges. Whether it's miles fitness and protection policies, zoning legal guidelines, or one-of-a-kind network bylaws, it's essential to be well-informed. Being proactive can save you quite a few time and effort. If you're handling regulatory objections, communicate with neighborhood government, be inclined to make critical adjustments, and make certain you are always compliant.

Internal Objections: Occasionally, the objections may additionally come from inside—be it from personnel, partners, or consumers. They ought to possibly question a modern method or the advent of a brand new product line. In such times, open communication is vital. Understand their concerns, back your picks with information, and be open to feedback.

Rejections, on the same time as associated, are a bit more definitive than objections. A belongings proprietor could in all likelihood flatly refuse a device, or a issuer won't need to offer you a particular product. In the ones instances, it's miles vital not to take it for my part. Business selections are frequently based mostly on a myriad of things. Always be professional, are looking for remarks on why the rejection happened, and use it as a analyzing possibility.